哈尔滨工业大学建筑学专业教学系列丛书

讨论·讲述·操作
Discussion　Illustration　Operation

哈尔滨工业大学与谢菲尔德大学
关于当代建筑教育的合作经验

Cooperative Architecture Education Program between
HIT and Sheffield University

徐洪澎 主编
李国友 邵郁 罗鹏 副主编

中国建筑工业出版社

图书在版编目（CIP）数据

讨论·讲述·操作　哈尔滨工业大学与谢菲尔德大学关于当代建筑教育的合作经验／徐洪澎主编. —北京：中国建筑工业出版社，2013.10
（哈尔滨工业大学建筑学专业教学系列丛书）
ISBN 978-7-112-15952-9

Ⅰ.①讨… Ⅱ.①徐… Ⅲ.①建筑－教育－国际合作－研究－哈尔滨市、谢菲尔德 Ⅳ.①TU-4

中国版本图书馆CIP数据核字（2013）第235856号

责任编辑：张　建
责任校对：肖　剑　刘　钰

哈尔滨工业大学建筑学专业教学系列丛书

讨论·讲述·操作

哈尔滨工业大学与谢菲尔德大学关于当代建筑教育的合作经验

徐洪澎　主编
李国友　邵郁　罗鹏　副主编

*
中国建筑工业出版社出版、发行（北京西郊百万庄）
各地新华书店、建筑书店经销
北京锋尚制版有限公司制版
北京顺诚彩色印刷有限公司印刷
*
开本：787×1092毫米　1/20　印张：13⅗　字数：630千字
2013年10月第一版　2013年10月第一次印刷
定价：88.00元
ISBN 978-7-112-15952-9
（24739）

版权所有　翻印必究
如有印装质量问题，可寄本社退换
（邮政编码100037）

丛书编委会

主　任
梅洪元

副主任
刘大平　孙　澄　金　虹　冷　红

委　员
徐洪澎　李国友　邵　郁　罗　鹏　殷　青
梁　静　吴健梅　韩衍军　王　岩

本书参编人员
（按姓氏笔画排序）

教　师
卫大可　王　岩　史立刚　白晓鹏　庄　葳
刘　滢　连　菲　吴健梅　张姗姗　陆诗亮
陈　旸　孟　琪　梁　静　董　宇　董健菲
　　　　　　　　　　　　韩衍军　蒋伊琳

学　生
丁凤鸣　刁　哲　李　思　邱　麟　张天硕

序一 Preface I

哈尔滨工业大学建筑学院有着九十余载的悠久历史，是中国最早建立建筑学专业的院校之一。作为当时远东最著名的教育平台，培育出了众多活跃在建筑领域的杰出人才。由于起源于中俄合作的教育发展背景，开放式教学和国际化教育环境成了这所高等学府突出的教育特色。随着历史积淀，其教育、教学体系先后吸收了各个时期的不同思想与理念；而来自不同国家与拥有多元教育背景的学者们在这里传授前卫的建筑思想，追逐世界建筑潮流，推动着哈尔滨这座当时北中国最开放的国际大都市的近现代转型和发展。

一个世纪过去了，教育在世界范围内发生了深刻的变化。中国建筑教育伴随着时代的脚步，面向世界，已在全球范围内进行着广泛而深入的交流。哈尔滨工业大学建筑学院地处容易使人感觉寒冷而封闭的严寒之地，区域限制常常成为发展的桎梏。然而，劣势在一定情况下也可以转变为优势：寒冷的气候条件促使我们深入研究寒地建筑创作和相关技术，"寒地"贯穿了科研与创作的各个方向，从而成为我们的专长与特色。而封闭的环境促使我们更加渴望交流，采取积极的态度去面对外界，用谦虚与包容的胸怀去接纳不同的观念与思想，同时努力为教师和学生创造更多共享世界建筑教育资源的机会。因此，当2010年我来到建筑学院的时候，心中就升起两个强烈的愿望：虽然地处气候寒冷的北方，但我希望我们始终保持对教育事业的热情和对教学工作的激情；我更希望学院在传承原有教育特色的基础上，更加开放和国际化，快速参与到国际交流的主流中来。

我的同事中有康健教授这样有责任感和影响力的学者，在他的积极努力下，2010年，"哈尔滨工业大学—谢菲尔德大学"中英建筑科学海外学术合作基地正式成立，这标志着哈尔滨工业大学建筑学院和英国谢菲尔德大学建筑学院的交流合作进入了一个更加广泛和更具高度的层面。随着友谊的加深、合作的深入，两院师生频繁往来，从两院联合举办中英建筑教育国际论坛，到谢菲尔德大学教授们一年一度的短期教学访问，再到我院师生及教授团赴英国考察学习，在一次次的交流与互访中，收获了友谊，夯实了合作，更促进了彼此在教育研究方面的发展。可以说，当今天回首这段愉快的合作时，已是硕果累累、受益良多。因此，我们特别制作了这本汇集着2010~2012年哈尔滨工业大学建筑学院和谢菲尔德大学建筑学院教育合作丰厚成果的专辑，以此见证两院深厚的友谊。即将出版之际，我谨以一个教育工作者的身份表示祝贺，并向那些在此期间付出了辛勤汗水的两院同行们表示感谢！

哈尔滨工业大学建筑学院院长、教授
2013年9月

序二 Preface II

　　教育应该是一个不断创新的过程，随着环境和社会需求的变化，建筑教育的理念和方法也应当不断捕捉更新、更高的目标。置身于当代社会，在我国快速城市化、经济全球化的浪潮中，世界越来越成为融合一体的舞台，开放应该是当代建筑教育院系保持生命力并不断创新的基本理念。在开放的状态下各种学术思想交锋论战，在开放的进程中校内与校外交流发展、学校与社会间互相促进。

　　作为英国谢菲尔德大学和哈尔滨工业大学之间的桥梁，我很高兴看到2010年"哈尔滨工业大学——英国谢菲尔德大学中英建筑科学海外学术合作基地"的建立，从此启动了两校建筑学院密切的学术与教学往来。三年间合作基地已在谢菲尔德大学培养哈尔滨工业大学青年教师和研究生数十人，并在哈尔滨工业大学成立了寒地建筑科学实验室。谢菲尔德大学亦有20余位教师到哈尔滨工业大学讲学交流。

　　每年的联合设计教学活动和建筑教学研讨会亦是两校合作的重要内容之一。以谢菲尔德大学专长的研究型教学与设计为核心，两校深入探讨了强调使用者、社会需求和可持续发展的教学理念及其在教学中的贯彻实施，为哈尔滨工业大学培养高水平的师资队伍，以及面向未来的学科发展等方面都发挥了重要作用。而在这一过程中，谢菲尔德大学的教师也对中国有了更为深入切实的了解，并为其教学带来了新的活力和发展机遇。

　　这本集结了2011、2012两年哈尔滨工业大学与谢菲尔德大学教学交流活动成果的书籍，承载了双方教师和学生教学、交流的累累硕果。从中可以明显看到该教学环节在整个建筑教学体系建设中所发挥的积极作用，即率先推进了哈尔滨工业大学建筑学本科生教学的国际化发展，而这个体系也正在日益充实和完善。学生在交流的过程中获得了批判能力、思维能力、实践能力、合作能力和表现能力的全面提升。

　　希望本书对国内关于建筑教育的探讨能够发挥抛砖引玉的借鉴作用。

康健

英国谢菲尔德大学建筑学院教授
2013年9月

前言 Foreword

当下，随着办学条件的提升和国际高校间沟通渠道的开辟，建筑学专业院校的国际化教学迎来了历史上最好的发展时期。2010年11月初，在哈尔滨工业大学长江学者、千人计划引进人才、英国谢菲尔德大学建筑学院康健教授的帮助下，谢菲尔德大学正式被列为获得"985工程"资助的哈尔滨工业大学海外学术基地，从此启动了两校建筑学院学术与教学的密切往来。三年多来，谢菲尔德大学来访教师达20余人次，两校举办建筑专业国际教学研讨会2次，组织国际联合教学workshop12组。在谢菲尔德大学进修半年以上时间的哈工大教师9人，参加短期访问交流的教师近20人。2011年9月在哈工大举办了"2011中英韩建筑教育国际论坛"，期间国内多所建筑院校教师与英国谢菲尔德大学、韩国汉阳大学的来访教师围绕信息时代的建筑教育主题展开讨论，同时举办多场学术讲座，并由谢菲尔德大学教师与哈工大青年教师一起组织高年级和低年级workshop各1组。2012年4月由吴健梅和罗鹏两位老师带领12名哈工大本科生前往谢菲尔德大学参加两校联合建筑设计。2012年8月末至9月初6位谢菲尔德大学教师到访，与法国拉维莱特建筑学院教授Eric Dubosc、西班牙建筑师Lorenzo Barrionuevo一起，面向我院2009级全体建筑学专业本科学生组织了5组联合建筑设计workshop，期间还举行了多场精彩的学术讲座。本书就是对2011、2012两年内哈工大与谢菲尔德大学有关建筑专业教学交流活动的总结，包括理论和联合课程实践两部分内容。

通过不同文化背景下的建筑教育理念的交流互动，我们感觉收获颇丰。首先，国外教学理念注重教师作为学生学习的引导者角色，鼓励学生大胆思考，并通过巧妙的方式引导学生按照正确的方法实现自我学习；而我们的教学，则更偏重与实践接轨，学生在设计创新思维及逻辑构思方面的能力相对薄弱，通过交流可以取长补短。其次，谢菲尔德大学建筑专业教学的特点是研究型设计题目的设置，无论从workshop的题目选择，还是所采用的教学方法都让我们深受启发，这种启发应与国内更多从事建筑教育的同行们分享。由此，我们产生了将两校教学活动成果总结出书的想法，以期抛砖引玉，让更多的师生获益。

本书的出版凝结了很多人的心血。如果没有康健老师的牵线搭桥，没有学校和学院的大力支持，没有谢菲尔德大学众多教师的积极配合，双方的深入交流与合作就无从谈起。此外，每一次交流活动都倾注了哈工大众多师生的精力和时间，这是远远超出正常教学工作量的；本书的编写老师和学生们做了大量认真、细致的工作，不厌其烦地进行调整和改动；建工出版社的张建编辑在本书的出版过程中提出了多项好的建议，并为本书在较短的时间内出版做了大量工作。这些都是我们要衷心铭记的，在此一并感谢大家！

2013年9月，哈尔滨工业大学与谢菲尔德大学的教学交流精彩继续，我们也期待在积累本书编写经验的基础上，不久的将来我们能有更深入的思考总结呈现给全国广大的建筑专业师生。

目录 Contents

教学论文——讨论
The Teaching Paper—Discussion

阶段控制　多向交流——关于建筑设计课程加强交流锻炼的研究	002
基于代码的建筑设计工作室	005
哈尔滨工业大学建筑学专业本科培养方案的优化	013
建筑设计基础教学的思路和方法探索	019
建筑设计教育探索——以汉阳大学安山校区为例	022
基于教学体系下的住宅建筑创新理念研究	026
传媒时代的建筑设计工作室教学——与不确定的未来一起工作	031
基于建筑模型的教学思考	042
传媒时代背景下的建筑设计教学课程	046
建筑教育国际化培养模式的理论与实践探索	049
开放式建筑教育的深化拓展	053
建筑设计系列课程开放式评价体制改革探索	057
将科学与技术融入设计教学	061
培养卓越建筑师的实践与创新研究	068
职业技能的重新思考："更加绿色的建筑师"	071
三个作品、一个人物引发的建筑教育思考	078
拥抱积极的灰色：揭示建筑理念	083

教学讲座——**讲述**
The Teaching Lecture—**Illustration**

基于大学的推动研究——以波特兰项目工程为例	092
数字化装饰	097
城市开放空间：城市生活的重要性	100
如何做一名快乐的建筑师？	104
英国的生态住宅	107
建筑与环境声学：跨科学、工程、社会科学及艺术的探索	111

联合设计——**操作**
Joint Design—**Operation**

01	未来场景——哈尔滨的意外建构	119
02	拼贴记忆——快速表达	139
03	区域调研——探索伦敦南岸	153
04	"6"度——关乎哈尔滨未来的设计	167
05	"下午茶"——学术活动	185
06	巴黎人居桥	199
07	哈尔滨国际象棋俱乐部	219
08	从具体到抽象，从抽象到具体	239

教学论文

The Teaching Paper

讨论 Discussion

阶段控制　多向交流
——关于建筑设计课程加强交流锻炼的研究 [1]

Stage Control Multi-directional Communication
—The Communication Exercise Reinforcement in the Architecture Design Course

李玲玲　Li Lingling
程　征　Cheng Zheng
哈尔滨工业大学建筑学院
School of Architecture, HIT

摘　要　本文针对哈尔滨工业大学建筑学院在建筑设计课程中加强交流锻炼的教学改革进行研究，强调设计课程的阶段控制与多向交流。对课程改革重点中期检查的教学反馈情况采用调查问卷的形式进行调查，为教学改革的进一步完善提供依据。
关键词　阶段控制，多向交流，教学改革

Abstract The paper researches the education reform of reinforcing communication exercise in the architectural design course.The stage control and multi-directional communication is the key of the reform.The teaching feedback of the mid-term examination is investigated in the form of questionaire,which provides the evidence (basic data) for the optimization of the teaching reform.
Keywords Stage Control, Multi-directional Communication, Educational Reform

当今建筑设计日趋成为一项更具有系统性和综合性的复杂工作。传统的建筑设计课程仅仅关注学生基本能力的培养，学生画图、老师指导、评判，这种简单的线性交流容易将学生们引入足不出户、闭门造车的死胡同，不善交流变为不愿交流，走上工作岗位后往往暴露出团队协作能力差、方案表达能力差等问题。

交流性教学是以信息交流为基本形式的教学，它包括教学过程中师生之间、学生之间、教师之间、人与书本之间、人与社会之间广泛的多形式、多渠道的信息交流。交流性教学与传统的传授式教学相比，前者信息是双向或多向传递，而后者信息仅为单向传授，缺乏交流。在设计课程的教学中加强交流能力的培养可以成为传统教学模式的有效补充。

1 多向交流的阶段性体现

设计课程一般分为四个阶段，即设计的准备阶段、设计的构思阶段、设计的完善阶段和设计的表达阶段，建筑设计课程交流化改革是建立在传统的阶段性教学基础上，针对实际调研、方案构思、方案讨论、中期检查等课程环节的不同特点，开展面向不同对象的方式多样的多种交流。交流性教学的分阶段、有区别的设计与控制成了课程改革的亮点。

1.1 教师引导、社会探索

设计的准备阶段由一节全体老师同学出席的教学大课作为开始，讲课老师按照不同年级的课程深度，为学生传授基本设计方法及设计思路，初步讲解任务书，安排课程进度与设计任务，这一环节是老师到学生的人与人之间的单向知识传授（图1）。为拉近学生与设计对象之间的距离，锻炼学生探索现实、发现问题的能力，在教学大课后安排学生组队进行社会调研，接近建筑，接近使用建筑的人群，感受建筑的功能、流线及形象特点，体验建筑使用过程中的种种现象，发现建筑实际使用中的问题（图2）。社会调研将学生拉出校园，接触现实，

---

1 本文为"2011中英韩建筑教育国际论坛"宣读论文。

避免空想，通过与社会大环境的交流，发现社会问题，激发设计热情，调研过程中的所见、所闻、所想往往成为设计的很好的出发点。

1.2 独立思考、班级讨论

设计的构思阶段是设计方案形成的关键环节，也是历时最长、交流最丰富的阶段。方案构思与方案讨论一静一动，独立思考与集体讨论相互穿插、相辅相成，制造更多的交流机会。随着网络时代的到来，知识的拓展不再局限于书本，学生与网络世界的交流更为直接、便捷。网络知识库的庞大使得学生与网络的交流越来越频繁，学生展示方案的手段也由过去的纸质拓展到了多媒体交流，用图片、模型甚至视频动画将设计思想表达得更加淋漓尽致，使交流更加充分（图3）。

老师与学生一对一的方案讨论可对方案进行比较详尽、细致的讨论，而班级内部的集体讨论则会引发更大范围内的多向交流。交流现场的所有人都在无形中被联系起来，一个问题可以引发最大程度的讨论，问题的解决方法也以最大程度传播开来（图4）。另外，公开讲解方案可以锻炼学生表达方案的能力与展示方案的信心。方案的陈列促使学生相互了解，取长补短，增强审美素养。交流是堆土成山、积水成河的过程，一个问题往往得出多种解答，再通过对比寻求最佳解决方案。

2 阶段控制的集中体现

2.1 中期检查的设置必要

经过设计的准备、构思阶段之后，大部分学生有了一定的设计成果，但这时往往也会出现一些问题。比如一部分学生由于外界信息刺激过多，无法抓住最强烈的设计意图，致使设计陷入混乱，对成果总不满意；二是有些学生设计理念新颖别致，但设计手法有限，造成方案总是停留在想法阶段，缺乏执行能力；另外，还有部分学生有了一定的设计成果，但在细化深入方面遇到瓶颈，无法使建筑更出彩。这时候，设置阶段性成果检测的环节，集中解决学生的疑难问题，帮助学生度过瓶颈期，为方案的最后冲刺排忧加油是十分必要的。经过教研室老师们的讨论，决定抛弃以往仅仅上交平面图纸作为阶段性成果的评判方式，增设中期检查环节，二维图纸、三维模型和学生讲解三者合一，使学生方案展示更为充分，成果评判更为公正。

2.2 别开生面的中期检查

由于是阶段性检查，为避免图纸表达占用学生太多精力，中期检查采用灵活的评判方式：每人给定50mm宽但长度不限的展位，学生可利用草图、草模等手段进行方案展示。为增强中期检查的全面性，在有限时

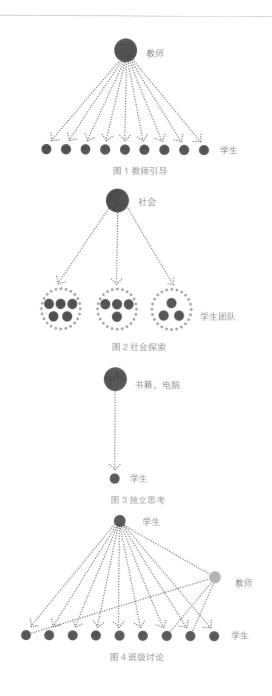

图1 教师引导

图2 社会探索

图3 独立思考

图4 班级讨论

间内更多地帮助学生解决问题，将中期检查分为两步：第一个环节是学生站在自己的展位前，老师快速简略地浏览学生展品，随时向学生发问；环节二：按照班级规模从每班挑选出3~5个典型作业，由作者在所有老师

和同学面前公开讲解方案，老师针对方案出现的问题进行提问及评价。

2.3 最为集中的多向交流

中期检查的交流范围扩大到了整个年级，每个学生的方案都面临着多位老师和全体学生的评判，是整个设计课程期间交流最频繁、最激烈、最充分的阶段。通过相互观摩、交流，发现自己的不足，获得前进的动力，为后续的设计完善阶段明确了方向。在中期检查的第一个环节中，每个方案都会接收到每位老师的信息反馈，同学们也可以看到整个年级的全部方案，接收的信息量大且具有针对性。整个年级所有的同学和老师之间都可能被交流的"红线"牵引起来（图5），帮助学生拓宽视野，突破瓶颈，撞击出意想不到的灵感火花，同时也有助于老师们对教学效果的宏观把握。如果说第一环节是多对多的讨论，第二环节则是一对多的交流，虽不是每位同学都有机会讲解方案，但是老师意味深长而又妙趣横生的点评也使学生们受益匪浅。典型方案的一些问题也会引起其他同学的共鸣，虽是针对性很强的评价，却可得到范围更广的传播（图6）。

3 多向交流的教学反馈

由于中期检查是新增教学环节，对学生影响的实际程度及学生的

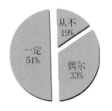

作业观摩环节的参与情况　　公开点评环节的参与情况

作业观摩环节对学生的帮助　　公开点评环节对学生的帮助

图7 中期检查情况调查

真实评价尚无一手资料。作者采用调查问卷的形式在二年级学生中随机抽取了36名，分别对作业观摩环节的参与情况、公开点评环节的参与情况、作业观摩环节对学生的帮助、公开点评环节对学生的帮助等问题进行了调查，调查结果如图7所示。

中期检查是为加强设计课教学交流而新增设的环节，第一次实行就收到了可喜的效果，在今后的教学过程中值得坚持和推广。

4 结语

实际工作中，交流是建筑设计师每日必修的课题。建筑师的自身技能固然重要，但只沉溺于自己的世界往往会导致思想狭隘、理念偏颇。在建筑设计教学中引入多向交流是对传统教学方式的重要革新。为师生提供一个直抒己见、畅所欲言的交流平台，不仅可使学生加深对建筑设计基本问题的理解，还可提升学生的自学能力、协作能力、创新能力，有利于正确工作方式的培养。在教学环节中增设中期检查是哈尔滨工业大学建筑学院的教学创新，也是为改善学生交流环境迈出的重要一步。

参考文献：

[1] 李建军，陈清.试论新形势下建筑教学"五环节"[J]. 高等建筑教育，1996(04).

[2] 李家莼. 交流：教学方法改革的突破口[J]. 高等教育研究，1999(02).

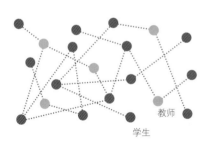

图5 作业观摩

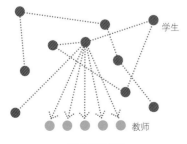

图6 公开点评

Understanding the Code-based Architectural Design Studio[1]
基于代码的建筑设计工作室

Mark Meagher[1], Thomas Favre-Bulle[2], Jeffrey Huang[2], Guillaume Labelle[2], Julien Nembrini[3], Trevor Patt[2], Simon Potier[2], Nathaniel Zuelzke[2]
1. School of Architecture, University of Sheffield, UK
2. Media and Design Lab, Ecole Polytechnique Fédérale de Lausanne, Switzerland
3. Chair of Structural Design and Technology, UdK Berlin, Germany
1. 英国谢菲尔德大学建筑学院
2. 媒体与设计实验室，瑞士洛桑联邦理工学院
3. 德国柏林艺术大学结构设计与技术学院

Abstract Traditionally, design learning in the architecture studio has taken place through a combination of individual work and joint projects. The introduction of code-based design practices in the design studio has altered this balance, introducing new models of joint authorship and new ways for individuals to contribute to co-authored projects. This paper presents a case study describing four design studios that used the computer as a tool for the auto-generation of architectural geometry. The format of the studios encouraged the students to reflect critically on their role as authors and to creatively address the multiple opportunities for shared authorship available with code-based production. The research question addressed in this study involved the role of code-based practices in altering the model of the design studio, in particular the understanding of authorship provided by visualizations of the digital design process.
Keywords Design Studio, Architectural Education, Code-based Design

摘 要 传统上来说，建筑工作室的设计学习通过独立设计和联合项目获得。基于代码的设计手法的引入在设计工作室中改变了这种平衡，并引进了新的联合设计的模型和个人参与合作项目的新途径。本文将展示对四个案例的研究，这四个设计工作室都是运用计算机自动生成建筑几何形体的。工作室的形式鼓励学生批判地看待自己作为设计者的角色，并创造性地把握基于代码生成提供的多种设计可能性。在这项研究中所涉及的问题有改变设计工作室模式、基于代码的实践研究，特别是在可视化的数字化设计过程中对建筑师角色的理解。
关键词 设计工作室，建筑教育，基于代码的设计

1 Introduction

Much has changed in design practices around the world as a result of the broad introduction of digital design tools. The use of design software has facilitated sharing of documents with collaborators in remote locations and the integration of analytical tools at multiple stages of the design process[1]. Code-based/algorithmic approaches to the design process have introduced methods for precisely controlling the outcomes of design and automating elements of the design process. Innovative construction processes have been enabled through the establishment of a direct link between digital data and the computer-controlled fabrication process[2].

Despite such transformative changes in the practice of architecture, the methods of design teaching have in most places been largely unaffected: following a formative period in the early 20th century, the basic principles of design education have remained comparatively constant. While the actual tools

---
1 本文为"2011中英韩建筑教育国际论坛"宣读论文。

used in the design process have changed, for example through the replacement of hand drawing with computer-aided design software, there have been few attempts to question the underlying presuppositions of design education in light of changes resulting from digital design practices.

The case study described here presents a three-year investigation of the potential for transformation of the design studio based on digital innovations in architectural practice. The project, known as 'Superstudio', was funded by the Swiss National Science Foundation and involved participants from three Swiss schools of architecture: EPFL, ETHZ and USI Mendrisio. Between 2008 and 2010, four design studios were offered in the Media and Design Lab at EPFL as a means of testing hypotheses developed by the Superstudio initiative. Students in the four studios were encouraged to work with code-based design processes, using computer programming to advance design ideas and to make constraint-based (parametric) simulation and fabrication a part of their design process.

Each studio in this series was organized as a means of testing the implications of integrating digital design methods in the studio. Among the implications considered were forms of communication enabled by the digital design process; the integration of bespoke parametric tools as an integral part of the design process; and the implications for design authorship of using code-based design methods.

1.1 Approaches to design education

The design studio is widely understood as the primary means of transmitting knowledge in the design professions. Many of the studio's defining characteristics are inherited from one of two historical sources: the 'atelier' tradition of the 19th century Ecole des Beaux Arts in Paris and the workshops of the Bauhaus, introduced in the 1920's.

The design education of the Ecole des Beaux Arts centered on the 'ateliers' or studios, rented spaces outside the formal context of the Ecole in which a group of students would conduct their design work assisted by an instructor or 'maitre' who provided regular critiques of work in progress [3]. The intellectual life of the ateliers consisted of constant interaction around design ideas. Over the course of the 20th century the atelier model emerged as the preferred method of design education in schools throughout the world, overtaking in most places the problem-solving method of the scientific disciplines. A significant evolutionary step in the history of the studio was introduced with the workshop-based teaching method of the Bauhaus in Weimar and the Vhutemas in Moscow. New techniques introduced by industrialization had fed a growing separation between the methods of craftsmen and techniques available through industrial production. The Bauhaus proposed to fill this gap through the reintroduction of a cultural practice within industrial development, combining both fine arts and the industrial practices. The Bauhaus workshop aimed to re-create the conditions of practice by according less importance to theoretical teaching than to the method of resolving a problem.

Although often based on 'real' problems and complexities, the design studio as a pedagogical instrument remains in most schools of architecture purposefully distinct from the practice of architecture and the other design professions [4]. The studio provides a risk-free environment in which students can safely try out ideas, learn the vocabulary and working methods of their profession, and learn design thinking through hands-on work and trial and error. The design studio introduces students to the process, materials and theory of design, creating an environment for learning that consciously deviates from that of the traditional classroom. In the studio, students learn to design by confronting complex problems using the tools, thought processes and practices specific to their profession. The studio master provides a model of professional and artistic excellence that students are encouraged to emulate, while much of the actual learning and refinement of skills takes place through peer-to-peer interaction [5].

1.2 The code–based design process

The Superstudio project focused on the generative uses of code in the design process and the implications of code for understanding contemporary changes in the atelier model of the design studio. Code was prioritized over other types of digital design tools such as CAD software packages because of perceived benefits in terms of both creative freedom and opportunities for collaboration.

The creative benefits of working with code include the high degree of control over the design process afforded by direct code-based production of geometry, and the relative transparency of constraints imposed by the software compared with CAD packages which often conceal from the user the assumptions and limitations inherent in the software [6]. Certain aspects of working with code also support collaborative work, or at least the possibility of collaboration. The modularity of well-structured code makes it relatively easy for students to share functional blocks of code, and the open source ethos of openly distributing and publishing one's ideas was actively promoted in the studios described here. And, the fact that students are using software with an international user community means that they can effectively draw on the internet to find code samples and to ask questions.

Through the use of code-based design techniques students were also able to access powerful libraries for environmental and physics simulation, applying performance criteria to their designs at an early stage in the design process. Through this integration of such powerful simulation techniques with generative design models, the studios investigated a shift in the role of the designer toward the control of an iterative design process in which automated generative mechanisms are used to achieve a meaningful interplay of form and function [7]. Another way in which the use of code-based design techniques can have significant implications for understanding the dynamics of the design studio is the capacity they introduce for detailed tracking of design decisions. The use of text-based computer files allows the integration of detailed version tracking in the design process, a system that provides a transparent record of authorship within a collaborative design process [8]. The visual representation of information describing the digital design process has the potential to encourage participation in collaborative design exercises and to facilitate the understanding of changing models of interaction within design teams [9]. Data collected during the process of design can be mined for implicit patterns using quantitative analytical tools, or visualized as a means of qualitatively identifying particular aspects of collaboration during the design process.

1.3 Visualizing the code–based design process

Information visualization is the branch of computer science research which deals with the communication of information in the digital medium. It involves the translation of data into a visual language, revealing inherent characteristics of the data that become legible as visually-recognizable patterns. As its name implies, information visualization focuses on visual representation of information, as distinct from other means of representing information such as sonorization. The field of information visualization is distinguished from scientific visualization by Card et al [10]. by its preoccupation with "abstract, nonphysically based data". Finally, information visualization is based on the use of computation to support data analysis, the generation of graphic output, and interactivity.

The visualization of data related to the code-based architectural design process has been explored in numerous research projects which have taken advantage of the activity logs generated whenever design work is produced digitally. By extracting relevant information from chat transcripts, email logs, version tracking software logs, and online collaboration tools such as wikis and blogs it can be possible to piece together a quantitative and qualitative understanding of the development of the code-based design process over time and models of authorship of projects produced by design teams. Beyond their use as a backup solution and file browser, the real importance of online collaboration platforms for the design process may lie in the insights they will eventually offer concerning the nature of collaboration and authorship in the digital design process. Among the precedent projects considered in the current study were Eventspace and OpenD, visualizations of the code-based design process developed at ETH Zurich and Harvard Design School.

Eventspace is a node-based collaborative visualization of design activities which provides members of design teams with the ability to publish their work online, comment on the work of others, and trace the chronology of individual contributions to collaborative design projects [11]. The system was first developed for use at ETH Zurich (Eventspace I) and was further adapted at the Harvard Design School where it has been used in numerous classes. Eventspace enables designers to place their contributions in the larger context of a digital archive, i.e., precisely situate where their own contributions fit in to the collaborative design process. By making it easier to identify the specific contribution of each individual, the visualization aim to create an increased understanding of the collaborative design process, an understanding that can lead to the identification of shared interests within a large group of designers thereby encouraging joint projects [12].

OpenD is a visualization of activity in an online shared workspace that focuses on presenting a global overview of interaction among the members of a design team; the software was developed at the Harvard Design School where it was tested in several classes [13]. The platform represents submissions to the online shared workspace as nodes which are connected based on tacit or explicit affiliation in terms of content and authorship. There are three basic visualization options: nodes organized by time (chronological genealogy of ideas), by people (who contributed what), and by keywords (as specified by contributors) [14]. The time organization provides a chronology of the emerging knowledge repository; the people view is a graphic representation that clearly identifies the contributions of each individual; and the keywords organization structures the knowledge base according to topics specified by contributors. At any stage in the design process it is completely transparent who has contributed what to a given design output, and which ideas have been most influential in the shared knowledge base.

2 A studio case study: Organicité s studio at EPFL

Between 2008 and 2010, a series of four architectural design studios were offered by the Media and Design Lab at the Ecole Polytechnique Fédérale de Lausanne (EPFL) as an experiment in the integration of digital design methods in the design studio and as a tool for studying the changes taking place in the structure of the studio as a result of this integration. In particular, these studios have been used a means of investigating the models of authorship in the design studio and the ways in which these models are changing in response to a code-based design process.

2.1 Anar+: Bespoke digital tools in the design process

A new coding library for architects, 'anar+', was developed by Guillaume Labelle and Julien Nembrini as part of the Superstudio project with the goal of creating a design tool sufficiently powerful for expert use while sufficiently forgiving for introduction to beginners [15]. The anar+ library for Processing was introduced to students in each of the four EPFL design studios, and was used to address particular design questions at discrete moments in the design process. Unlike monolithic software packages such as AutoCAD which attempt to provide a one-size-fits-all solution to the varied needs of its users, anar+ was designed as a means of enabling users to create their own tools to address particular design questions at a specific stage in the design process.

The anar+ library supports the parametric definition of 3D

geometry using constraints based on geometric constructs, performance, or other factors defined by the user. Once the geometry has been created, the anar+ graphical interface allows continuous variation of parameters in real time using sliders. anar+ is designed to be used in conjunction with software for 3D modeling, 2D drafting, and rendering, and includes export capability to common design software file formats such as Wavefront OBJ, Rhino, Sketchup, and AutoCAD.

To give one example of the way in which anar+ was integrated in the design process, in the Organicités spring 2009 studio the anar+ library was used in the early stage of design to generate concepts for massing and interior plan organization of a skyscraper. The building envelope and floor plans were generated based on a range of programme-related and environmental performance criteria which were researched and implemented as geometric constraints by the students. Following the development of these overall concepts for the massing and interior the model was exported and the detail design stage was later carried out using one of several standard 3D modeling software packages (Fig. 1).

The students in the Organicités studios were encouraged to share the code produced using anar+ in an online shared workplace set up for this purpose. Questions about particular aspects of using the software could be addressed with peers and instructors using this online interface, which was conceived and implemented by Guillaume Labelle and consisted of an integrated platform of blogs, wikis, and a 'subversion' (SVN) version tracking system. The online platform was intended as a means of encouraging collaboration among the students: by providing a means for students to share their work and a visual interface for browsing the work of others it was expected that students would more readily find others with shared interests or technical challenges with whom to collaborate. The organization of the studio also encouraged students to form ad-hoc groups for the completion of a particular task or for the production of a particular design concept.

In addition to this emphasis on group work, an 'open source' ethos of sharing one's work and learning from the work of others was promoted in all the Organicités studios. Students were encouraged to post their code online at multiple stages in the design process, and to download and use code produced by their peers. This borrowing of code was tracked using a version control software which logged the 'use' of any given file in the online repository and kept track of the

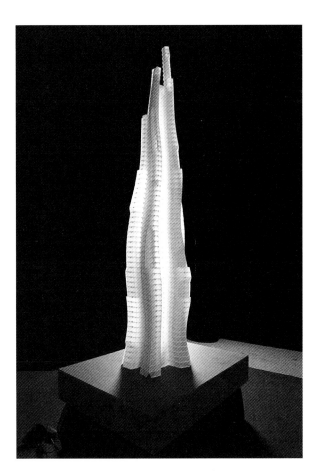

Fig. 1 Organicités skyscraper model by Osamu Moser.

authorship of code on a line-by-line basis, providing a detailed record of who had contributed to a jointly-authored work. In this way, a record of individual authorship was preserved within the collaborative design process.

2.2 Functions of the online repository

As mentioned above, the online repository served multiple functions related to facilitating the exchange of digital design knowledge. One frequent and unfortunate side effect of the digital design studio is a decline in the output of physical objects, a fact that makes it quite important to provide a digital repository of design work that students can easily browse and explore. The two primary purposes of the Superstudio online repository was 1) to allow students to share their work with their peers, and 2) to keep track of individual authorship when student design work was produced collaboratively.

First, the online repository provided a visual interface for students to 'publish' their code and design ideas. Students were encouraged to publish their work as frequently and as compellingly as possible using the online forum: this act of publication was both a record of authorship, a 'proof' of the origin of a particular idea with a given author or group of authors, and a means of soliciting feedback from peers and instructors. Essential to this 'explicit' form of information gathering were the comments students wrote to accompany their submissions to the online repository and the images they chose to represent their work.

The second goal of these online environments was to capture tacit information about authorship. For the ethos of sharing and collaboration to remain compelling to students, there had to be a mechanism in place for keeping track of individual contributions to group projects with a high level of precision. This is much easier to do with code than with traditional design practice–because a significant element of design intention is encapsulated in modular units of code which can easily be tracked. This is what the version control system accomplishes– it keeps track on a line-by-line basis of who is the author of a particular segment of a given file.

2.3 Methods for capturing design data

In addition to its utility as a tool for communication in the studio, the Superstudio online interface was also used as a means of collecting data relevant to understanding the digital design process. Once an anar+ script was submitted to the online repository it became possible to track who had viewed, downloaded and contributed to the script. Because students were encouraged to learn from and actively build on and use the work of their peers, there was considerable activity of this kind.

In the Superstudio project we tested multiple tools for collecting data related to the digital design process. In the first two iterations of the EPFL studio students were required to submit their digital files to an online file server each week, and these files were then made available to all the students on a web site. In the third and fourth studio iterations 'subversion' (SVN), a powerful open source software commonly used by computer programmers, was implemented by Guillaume Labelle as a means of tracking changes made by multiple developers to a shared code repository. SVN maintains a backup copy of the current version of every file submitted by the students, and also of all previous versions, allowing students to quickly access any stage in the design process. SVN also stores each change to the files–among the data recorded is authorship, timestamp, and a record of precisely what changes were made.

A visualization of the SVN log was designed and implemented by Thomas Favre-Bulle and Simon Potier, and this was used in two iterations of the Organicités studio series. This visualization is shown in Fig. 2: the diagram describes the authorship of a project produced by a number of different students over the course of the semester. Each horizontal line represents an individual file; the colors represent each author

Fig. 2 Visualization of the SVN file repository developed by Thomas Favre–Bulle and Simon Potier.

who contributed to the file. As the visualization indicates, some files are individually authored while in a number of cases there are multiple authors; the profile images for each of the authors is displayed on the right. It is possible for students to click on each section of the line, and see exactly what lines of code were written by whom.

3 Discussion and future work

In future implementations of the Superstudio online repository it will be important to gather data related to the influence of the visualization itself on the students and their own understanding of the digital design process. In the studios discussed here the visualization was used as a means of presenting students with a visual representation of certain aspects of their own digital design process, and it will be interesting to determine whether the availability of this visual representation has a significant impact on the students and their understanding of their design process.

Among the primary challenges in the visualization of the digital design activity is the capture of meaningful information during the design process. Tacit methods of information capture such as data mining of server logs, version control software logs, email logs and chat transcripts can be quite effective and reliable, and this has been the preferred method in the examples cited above. The limitation of tacit methods is the nature of the information captured in logs, which seldom provides insight into the motivation and thought process behind a given design idea. For this reason tacit methods can sometimes lead to misleading or overly simplistic conclusions due to the absence of a context for interpreting results. Among the challenges to be addressed in future studies is the development of a better method for collecting meaningful information about the design process through a combination of tacit and explicit data collection methods.

References:

[1] Nembrini J, Labelle G & Nytsch-Geusen C. Parametric scripting for early design building simulation. CISBAT, 2011.

[2] Shah J, Mäntylä M. Parametric and feature-based CAD/CAM: Concepts, techniques, and applications. Wiley, 1995.

[3] Drexler A. The architecture of the ecole des beaux-arts. Cambridge, MA: MIT Press, 1977.

[4] Findeli A. Rethinking design education for the 21st century: Theoretical, methodological, and ethical discussion. Design Issues, Vol. 17, No. 1. (Winter, 2001), pp. 5-17.

[5] Schön D A. The architectural studio as an exemplar of education for reflection-in-action. Journal of Architectural Education (1984), Vol. 38, No. 1. (Autumn, 1984), pp. 2-9.

[6] LaBelle G, Nembrini J & Huang J. Programming framework for architectural design [anar+]. In CAAD Futures'09, (Montreal, Canada), 2009.

[7] Nembrini J, Samberger S, Sternitzke A & LaBelle G. The potential of scripting interfaces for form and performance systemic co-design. Proceedings of Design Modeling Symposium. Springer, 2011.

[8] Rocker I. Versioning: Evolving architectures-dissolving identities: nothing is as persistent as change. Architectural Design vol. 72, iss. 5, p. 10, 2002.

[9] Meagher M, Bielaczyc B & Huang J. OpenD: Supporting parallel development of digital designs. Proceedings of ACM Design of User Experience Conference, DUX 2005, San Francisco, November 2005.

[10] Card S K, Mackinlay J & Shneiderman B. Readings in information visualization: Using vision to think. Morgan Kaufmann, 1999.

[11] Hirschberg U. Transparency in information architecture: Enabling large scale creative collaboration in architectural education over the internet. International Journal of Architectural Computing, vol. 1, iss. 1, p. 12- 22(11), 2003.

[12] Spicer D E, Huang J. Of gurus and godfathers: Learning design in the networked age. Education, Communication & Information, Routledge: Vol.1, No.3, 2001.

[13] Meagher M, Bielaczyc B & Huang J. OpenD: Supporting parallel development of digital designs. Proceedings of ACM Design of User Experience Conference, DUX 2005, San Francisco, November 2005.

[14] Huang J, Bielaczyc K & Meagher M. Liquid ontologies, metaperspectives, and dynamic viewing of shared knowledge. Proceedings of I-KNOW 6th International conference on Knowledge Management, Graz, 2006.

[15] LaBelle G, Nembrini J & Huang J. Programming framework for architectural design [anar+]. In CAAD Futures'09, Montreal, Canada, 2009.

哈尔滨工业大学建筑学专业本科培养方案的优化 [1]
The Optimization of Training Scheme for the Bachelor of Architecture in Harbin Institute of Technology

徐洪澎　Xu Hongpeng
李国友　Li Guoyou
哈尔滨工业大学建筑学院
School of Architecture, HIT

摘　要　面对社会人才需求的不断变化和建筑教育理念的快速发展，哈尔滨工业大学每四年会对现有培养方案进行一次优化。结合近年来的教改试验并以相关教学课题研究为基础，2012版培养方案的优化思路逐渐形成。新的培养方案力图进一步强化传统特色、推进国际化进程以及提升教学效率。本文是对新培养方案的全面审视和总结，以吸纳多方建议，推动相互借鉴，促进成果完善。
关键词　建筑学专业，教学体系，实践，理念

Abstract　Because of changing demand of talents and rapid development of architectural education concept, the existing training scheme in Harbin Institute of Technology is optimized every four years. Being combined with the experiment of educational reform and based on the relevant subject study, the idea of optimization of 2012 training scheme is gradually developing.The aim of new training scheme optimization is to strengthen traditional characteristics, boost the international progress and improve the teaching efficiency.The paper absorbs collective wisdom, promote mutual reference and improve the results so as to conclude and survey the new optimization more comprehensively.
Keywords　Architecture, Teaching System, Practice, Concept

　　21世纪以来，在信息化环境下的建筑专业发展是革命性的：争奇斗艳的设计思想、不断突破的建筑技术和高科技辅助设计手段严重冲击了课堂上教与学的传统模式；"授人以渔"的教学理念促使传统授课模式向以学生为主角的现代教学方式快速转变；教育视野的扩展和教学资金的注入为国际联合设计等更灵活的课程形式的出现奠定了基础……这些变革推动了建筑院校对自身教学理念与教学体系的重新审视。同时，社会对建筑专业人才需求的剧增催生了大陆地区建筑学专业院校数量的迅速增加，其总数已经由20世纪的不足80所增长到现在的252所。如何在同大于异的全国建筑教育系统中定位自身角色并细化专业培养目标，突出教育特色，提高教学效率，发挥人才培养优势，这些都已经成为各个学校在教学体系优化中所面临的重要课题。

　　在此背景下，哈尔滨工业大学建筑学专业于两年前开始准备对教学体系进行优化，并在教学实践中进行了一系列的教改尝试。经过研究我们认为，对于已有九十多年发展历史的专业来说，原有的教学体系积淀了较强的科学性与体系化特征。"新"教学体系方案不应是颠覆性的，而应在原体系上进行深化和优化。大框架坚持以专业设计课程为主线，建筑理论、建筑技术以及艺术类板块课程根据主线形成系统对接。根据目前的教育背景与教学资源等现实情况，将大结构调整为2+1+1+1的体系，即：第1、2年定位专业基础，第3、4年定位不同专业能力的扩展，最后一年为专业实践与毕业设计（图1），以此为依据进行课程选择和组合（图2）。课程在总体培养目标的基础上，进一步明确阶段目标和课程目标的三级目标体系，为具体教学工作形成更清晰的目标导引。新教学体系方案的优化思路归结为以下几项内容。

1 强化传统优势与特色

　　哈尔滨工业大学建筑专业的历史可以追溯到1920年俄国修建中东铁路时成立的哈尔滨工业学校铁路建筑科，当时这里的教学水平相对俄国本土来说也是高水准的，众多知名的俄国建筑师和学者曾在这里工作。"学院派"和职业学校的属性奠定了哈工大注重建筑创作基本功

---
[1] 本文为"2011中英韩建筑教育国际论坛"宣读论文。

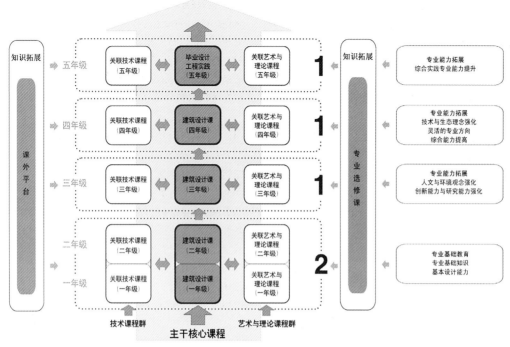

图1 哈尔滨工业大学建筑学专业课程体系方案框架图

年级		一年级			二年级			三年级			四年级			五年级	
		上	下	小	上	下	小	上	下	小	上	下	小	上	下
通识课		体育 外语 计算机 思想道德修养与法律基础	体育 外语 中国近现代史纲要		体育专项 外语 毛泽东思想和中国特色社会主义理论体系概论	体育专项 外语 马克思主义基本原理		体育专项	体育专项						
		文科数学	画法几何与阴影透视			建筑力学									
								人社基选1	人社基选1		人社基选1	人社基选1			
专业教育	专业必修	设计基础1	设计基础2		建筑设计1	建筑设计2 建筑设计3	快速培训	住区规划与景观 建筑设计4 快速1	建筑设计5 建筑设计6 快速2		开放设计 建筑设计7 快速3	建筑设计7	联合设计	业务实践	毕业设计
	艺术必修	造型基础1	造型基础2		艺术专题1	艺术专题2		艺术专题3	艺术专题4						
	技术必修				计算机应用	建筑材料 建筑构造-1		建筑结构选型 建筑结构 建筑构造-2			建筑设备（水暖电）	建筑物理（声光热）			
	理论必修	建筑概论 专业导论			中国建筑史1 公共建筑设计原理	城市住区规划原理		外国建筑史 住宅设计原理	中国建筑史2		城市设计概论 建筑安全性	建筑师业务实践培训			
	专业选修				选修2学分	选修2学分		选修3学分	选修3学分		选修5学分	选修5学分			
	跨专业				选修1学分	选修1学分		选修1学分	选修1学分		选修1学分	选修1学分			
实践类		军训及理论			绘画、实习1 表现实习1			绘画、实习2 表现实习2			测绘实习 构造实习		调研实习		
创新类1								2学分							
合计学分		25	19.5	5	21.5	22		22.5	21.5		22	21	5	14	14

图2 哈尔滨工业大学建筑学专业课程方案简表

与注重技术的传统教学特色,尤其在草图、渲染、二维尺度与比例控制以及建筑技术知识等方面的教学能力比较突出。新中国成立后,建筑专业的师资构成中技术专业教师占据很大比例,比如梅季魁等知名教授都是建筑技术专业出身,由此进一步强化了建筑技术力量强势的专业特色,并使哈尔滨建筑工程学院(从哈尔滨工业大学分离出来后学校的名称)很快在技术含量较高的体育馆、影剧院、高层建筑等的设计与教学领域占据全国至高位置。几十年来,哈尔滨工业大学毕业生相对坚实的专业基本功和处理建筑技术问题的能力受到用人单位的普遍认可。此外,依托地域条件的哈工大一直将寒地建筑创作及其理论作为其重点科研方向,在建筑专业全部9个研究所中有3个从事此方向的研究,经过几十年的积淀,凭借黑龙江省"寒地建筑重点实验室"等优越条件,寒地建筑已经成为哈工大建筑特色和优势教学领域。

这些教学特色不但要推动继承,还要大力发扬。需要注意的问题是,建筑教育从19世纪初受巴黎国家艺术学院影响的传统建筑教育理念主导,到20~30年代的包豪斯建筑教学模式的全球蔓延,再到如今建筑教育所呈现的多元化趋势,发展传统教学特色如果僵化地固守原来的教学内容是不合时代要求的。经过研讨,我们确定了将草图、二维尺度与比例把握、三维空间建构和建筑创作的基本思维方法作为重点训练的基本建筑创作能力,将建筑结构选型、寒地建筑技术和计算机技术模拟等作为建筑技术的重点教学内容;而在寒地建筑创作教学领域,仍然坚持建筑空间、形态和技术的全方位知识覆盖。以此为依据,优化教学体系,将建筑创作基本功的训练主要集中在前两年,并在后三年的课程中进一步强化。建筑技术能力主要通过高年级课程加强,不但坚持两个完整的共一百多个学时的综合设计课程,而且依照"改原理讲授为原理应用"的原则在高年级结合设计课程安排"嵌入式"建筑技术课程板块。2012年已经试点了技术教师参与设计课程指导、并要求学生完成单独技术报告的教学方式。在寒地建筑课程方面则更多依靠学科的深厚科研基础,提供相关选修课程,并结合综合建筑设计等主干课程进行专项训练。

2 推动国际化教学发展

尽管哈工大建筑学专业的建立在一开始就具有国际化的特色,但是由于后期国际形势的变化及地处经济与文化不发达地区的不利影响,在很长一段时间内,我们的专业教学环境是相对封闭的。在当前全球化的背景下,国际化交流已经形成趋势,也成为了卓越建筑专业人才培养不可或缺的环节。为此,近年来学校、学院大力发展教学的国际化建设,据不完全统计,2010~2012年的三年中建筑专业开展国际教学交流活动百余次,包括国际教学会议、讲座、联合设计工作坊、外教联合教学、座谈以及出访等多种形式,来访外国学者和学生500多人次。到目前为止,与哈工大建筑学院交流的学校有美国的麻省理工学院、哈佛大学、加州大学伯克利分校、英国的谢菲尔德大学、荷兰的代尔夫特理工大学、德国的柏林大学、意大利的都灵理工大学、法国的拉维莱特大学、俄罗斯的莫斯科大学、韩国的汉阳大学以及台湾地区和香港特区的多所大学。

国际化建设让师生普遍感到收获颇丰:首先,国外灵活的教学题目、独特的教学方法、前沿的教学内容极大拓展了师生的知识储备,尤其在建筑理论方面的教学与研究让我们看到了极大的差距;其次,近距离的接触增加了对国外文化的了解,增进了友谊,建立了长久的交流渠道;再次,学生在交流中积累了信心,每年近1/3的毕业生出国深造也得益于此。在多种交流模式中效果最好的是时间相对较长的联合教学等方式,能够保障交流的深度,这也是新教学体系方案重点搭建的国际化教学平台。在5年的学业中,安排每名学生至少有一次国际联合设计课程的经历,至少参与2门由外国专家讲授的理论或技术选修课程,至少有一半的同学能够获得到大陆以外地区进行教学交流或交换培养的机会。确定于四年级开设的国际联合设计课程已经连续尝试了三年,每个设计小组都是由聘请的外教和中国助教联合指导,题目、教学环节由外教拟定,同样是全年级学生自由选择。尽管设计周期只有1~2星期的集中时间,但是教学过程和成果都比较理想(图3、图4),学生作业连续2年获得全国大学生优秀作业奖励。

3 促进课程的先进性与高效率

教学体系对课程的规定应在稳定与灵活性上找到最佳平衡点。规

图3 开放设计不同小组的学习过程

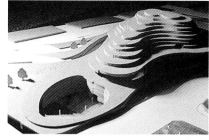
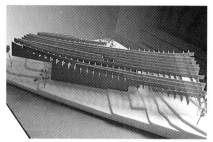

图 4 国际联合设计各小组设计过程与部分成果

定过少，课程有可能会游离于教学体系之外；规定过细，又会限制课程的及时更新。一些国外名校的课程，其教学目标、学时甚至课程表几十年没有变化，大大提高了教学管理效率，而主讲教师、教学内容、形式和方法却始终处于更新状态，一直保持着世界领先的地位。本次教学体系调整方案对课程的规定也只包括课时、学分、目标与核心知识点这4项内容，其他由课程小组自定以增加课程建设的灵活性和先进性。比如将"城市环境群体空间"课程的名称改为更宽泛的"建筑设计6"，明确其教学目标为树立城市环境建筑创作观念，要求学生以研究的方法解决城市建筑的设计问题。由于减少了束缚，在这两年的实践尝试中，课程教学小组不断进行题目、内容、方法等的教

大跨度建筑与结构综合创新设计

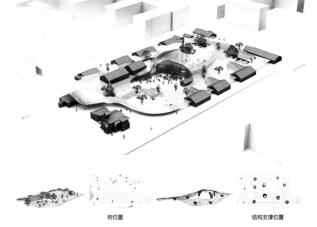
历史街区保护与复兴

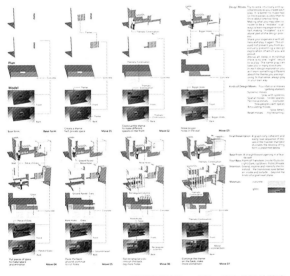
研究建筑产生、成长、改变的设计游戏

参数化建筑设计

图 5 开放设计不同小组的部分成果

学改革。类似的教改也发生在其他课程中，学校在2年的全国高等学校建筑设计优秀教案和教学成果评选中有5门课程获得优秀教案奖励也从中获益（图5、图6）。

提高教学效率的重点之一是主干设计课程的设定。一种举措是增设了小课时的专项训练。以往每一个课程设计都是相同的课时、类似的过程和平、立、剖外加效果图的相似表达，学习内容重复，而专项设计可以不完整、减少环节、局部表达，用更短的时间对某一问题形成深入的学习。另一举措是整合了多科目的综合训练课程。比如将城市设计、建筑设计和室内设计三个部分的内容打包在一个大设计课程中，先以一个题目做城市设计，然后在城市设计成果的基础上做建筑设计，最后在建筑设计成果的基础上再做室内设计。课程贯穿整个学期，相对以往的单独设置，总学时虽然压缩了将近

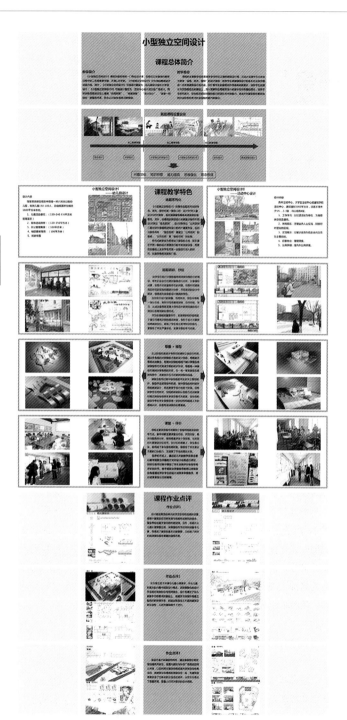

图 6 获得 2012 年全国高等学校建筑设计优秀教案和教学成果奖的教案

1/3，但是因为减少了调研等大量基础性工作，既缩短了学生了解题目的时间，又能使学生保持相对一致的专注度，课程深度将不受影响。同时，各部分的时间控制可以根据题目的复杂程度和教学环节等的实际情况进行实时的调整。由于当前建筑师经常会在一个项目中应对城市、政策、技术等综合问题，所以这种模式的课程也是对社会需求的适应。当然，这样的课程需要不同专业的教师很好地协调，增加了教学管理的复杂程度。

4 结语

哈尔滨工业大学建筑专业历史悠久、积淀深厚、生源和师资优秀、教学特色鲜明。对教学体系进行优化为的是适应时代需求、强化自身特色、提高教学效率。虽然地域偏远，但是这里的师生对教学十分投入和专注，这为教学体系优化奠定了重要基础。经过近两年的实践尝试与理论准备，新教学体系方案将于2013年正式执行。本文总结了在新教学体系方案制定过程中的实践体会和核心理念，力图在此方案执行之前梳理思路、深化总结，也借此机会征求同行意见，以便进一步完善。

参考文献：

[1] 朱文一. 当代中国建筑教育考察. 建筑学报，2010，10:1-4.

[2] 梅洪元，孙澄. 引智聚力特色办学——哈尔滨工业大学建筑教育新思维. 城市建筑. 2011, 03:27-29.

[3] 仲德昆，屠苏南. 新时期新发展——中国建筑教育的再思考. 建筑学报，2005，12:20-23.

[4] 常青. 建筑学教育体系改革的尝试——以同济建筑系教改为例. 建筑学报，2010，10:4-9.

[5] 吴健梅，徐洪澎，张伶伶. 中德建筑教育开放模式比较. 建筑学报，2008，07:85-87.

建筑设计基础教学的思路和方法探索 [1]
The Teaching Methods of Basis Architecture Design Education

李 伟　Li Wei
袁逸倩　Yuan Yiqian
天津大学建筑学院
School of Architecture, Tianjin University

摘 要 论文以国内各高校的教学改革实践为基础，以现代建筑基础教育要适应时代要求为准则，提出建筑基础教育应实现从"知识累加"到"思维建构"的方法转变，并在建筑设计基础课程改革中积极寻求提升综合思维能力的培养模式。
关键词 建筑教育改革，思维建构

Abstract The dissertation is on the basis of teaching reform practice in universities of China. As a guideline that modern architecture foundation education should adapt to the requirements of the times, the paper holds the opinion that the architecture foundation education should transform from "Knowledge accumulation" to "Thinking construction" and positively looks for the training mode of improving comprehensive thinking ability in the process of curriculum reform of basis Architectural design education.
Keywords Architecture Education Reform, Thinking Construction

1 建筑知识的累加和思维方式的建构

纵观国内各高校肇始于20世纪末的建筑设计基础课程的教学改革，迄今为止已十年有余。回顾多年来的教学改革历程，国内各高校建筑基础课程的教学内容和目标架构应该说主要由两个方面构成：

天津大学建筑设计基础课于2010年被评为全国精品课程。回顾13年来的教学改革历程，我们认为，天津大学建筑基础课程的教学内容和目标架构由两个方面构成：

第一，"建筑知识"——掌握建筑设计表达的方法和方式，其中包括了解建筑的基本设计规范、掌握建筑制图的相关规范、学会建筑设计的常规表达技巧等，即"建筑职业技能基础知识"。

第二，"思维建构"——体会建筑创作的理念和方法，并尝试和探索表达设计理念的多种设计手法和方式，即在发现问题的基础上，探索解决问题的多种途径，即"建筑设计专业思维素养"。

对于刚刚经过传统应试教育走进大学的学生来说，建筑学专业的基础教育不应仅仅停留在对建筑的一些基本知识和表达技巧的掌握，它应致力于使学生具有正确的建筑设计思维方式和设计方法，应该在学生学习建筑设计的初始阶段，培养他们对于所学专业的多向度思维，充分调动主体思维的能动性。因而，在20世纪末期的各高校建筑设计基础课程教学改革中，逐渐将课程的主要教学目标转变为在给学生传授建筑基础知识、技能的同时，更多地关注于学生建筑创作方法的思维建构。天津大学于1999年起实行教学改革后，我们将建筑设计基础课程的主要教学目标改变为：在给学生教授建筑基础知识技能的同时，更多地关注学生建筑创作方法的思维建构，并通过一系列的课程训练，使学生在学习知识框架上形成"建筑知识累加"与"思维方式累加"相互补充，在课程中逐步实现从"知识累加"到"思维建构"的教学内容的转变。

2 建筑设计基础课程教学思路和方法探索

国内各高校在建筑基础课程教学改革中大多遵循以"思维建构"为主的教学目标，归纳起来，改革后的建筑设计基础课程主要从以下

---

[1] 本文为"2011中英韩建筑教育国际论坛"宣读论文。

三个方面对教学思路和方法做了相应的探讨和实践。

2.1 拓延"基本功"概念——强调"思维设计技能"基本功的培养

基本功的练习和训练是各个专业必备的专业技能。传统"基本功"的定义是：使学生通过大量的相关练习，达到技能的纯熟性，因而在其进行更深层次专业行为的时候不会形成障碍。例如一个建筑师应该使自己的基本技能达到纯熟的程度，如绘图、勾草图等，而不会妨碍日后表达建筑设计理念的进程。

在我国现阶段的具体条件下，基本功的训练不可偏废，问题的关键在于我们应如何顺应时代的要求，给予"基本功"新的定位，拓展建筑基础教学"基本功"的概念。对于其他专业而言，基础阶段的教育往往采用灌输式的教学方式，学生掌握一些应知应会的基础知识，为将来的研究工作和实际操作做准备。而在建筑学中，我们知道技能和各专业基本知识是可以计量并可以教授的部分，甚至包括复杂艰深的高新技术和层出不穷的材料知识，我们可以称之为"硬"基本功的学习，但建筑设计基础阶段却不同，它包括大量不可计量部分的学习，这包括工作方法、创造能力、感知能力、理解能力等，可以称之为"软"基本功的学习。这种重视思维建构的培养方式对于刚刚经过传统应试教育走进大学的学生来说更为重要，这些恰恰是我们以前忽视的部分。"软"与"硬"、"虚"与"实"两者恰如其分地结合，才是一个完整的、真正意义上的建筑学。

因此，建筑学的基本功训练应把以"图面训练"和"知识累加"为主的传统"硬"基本功训练拓展为以注重创作主体"思维建构"为主的"软"基本功训练，即"感知空间的能力、驾驭空间的能力、创造空间的能力、动手实践的能力，和丰富的建筑语言表达能力"等。

例如，天津大学建筑学院是国内较早进行教学改革的院校之一。从1999年开始，通过多年不断的改革与探索，基本形成了一套比较适合现代建筑设计基础教学的框架，使一年级设计基础课的教学成绩逐年提升，学生创作出了一些独具风格的设计作品和成果，并多次在"全国大学生建筑设计作业观摩与评选"中获奖。其相应的课程作业设置如下：

（1）人体尺度的感知与认知——"满足一个人生活的集约化空间"

作业的基本内容是：在测量人体尺寸和与人体有关的家具、房屋门窗等尺寸的基础上，设计一个满足一个人生活的集约化空间。作业要求学生初步了解空间与人体尺度之间的关系，了解人体工学的基本知识，了解并运用基本的建筑构件，如楼板、墙、柱、门窗、坡道、台阶等作为形式构成要素（相当于点、线、面等基本要素），进行有目的、有意义的空间组合，建立起有序的空间秩序。

因为这是学生入学的第一个作业，重点要求学生学会做设计笔记，即：记录构思草图及设计过程，用PPT文件的形式总结汇报。

（2）城市尺度的感知与认知——"城市某重点地段的空间印象"

如果说上一个作业"人体尺度的感知与认知"是让学生在微观层次上对空间的人的行为进行认知的话，那么，这个作业就是在更加宏观的角度上，让学生通过自己的观察和城市体验对相应的空间尺度作出评判和提炼。

以"现场调研"作为设计题目的开端。通过调研工作切入生活实际，让学生有机会接触设计的前期工作，建立宏观的设计理念，理解建筑与城市的关系。使学生的设计方法由"书本上的设计"渐入"生活中的设计"，为将来面向社会，适应社会打下良好的基础。作业要求学生在调研的基础上自拟任务书。

作业要求用图片剪辑作为空间想象的一种手段。这种想象与"空间构成"作业中的想象是不同的。"空间构成"作业涉及的是一种抽象的空间想象，所凭借的是空间限定构件，如垂直和水平构件、线性和平面构件等，而"场景拼贴"所追求的不是一个抽象的空间，而是一个具象的场景。在这个作业中，照片剪辑是空间思维的一种辅助工具，其关键并不在于画面本身，而在于所创造的空间的品质。设计中要求学生体现"建构"的特点，即用一些照片的片断来"构筑"空间。总之，空间想象是这个课程设计练习的基本目的，也是我们老师评价作品的主要标准，关键是看作者是否能给观者带来不同的空间体验。此外，这个作业还可以锻炼学生对建筑尺度的把握与认识，也可以说，这是一个综合性的练习。

2.2 "思维建构"设计方法的全面引导——从空间感知到空间认知

在全球化的今天，对于初入大学、开始接受专业教育的学生们，究竟应该选择怎样一条道路呢？基于"思维建构"这个中心，我们可以把设计能力的培养分解为三个层面，即捕捉问题的敏锐力、分析问题的洞察力、解决问题的创造力，并把其分解成六大思维能力的建构，即感知力、观察力、分析力、理解力、想象力和表达力。

很多高校在这个环节中，以空间系列设计作为建筑基础训练课程的重点内容，并将这六大思维能力的培养全面贯穿于"空间生成"的系列训练中。通过一系列由简单到复杂、由概念性空间到实用性空间的循序渐进的创造过程，将纯粹的空间设计纳入建筑专业训练体系，遵循从概念到形式、从三维空间到二维平面、从模型设计到图纸设计的认知过程，培养学生的空间想象能力和复杂空间的营造能力，借助实体建构培养学生对抽象空间的认知能力，对空间尺度的控制和设计能力和对空间的构造节点、技术材料的掌控能力。旨在从素质教育的角度出发，根据建筑基础教学的特点，通过空间感知到认知的研究和实践，探索学生如何最大限度地开发和建立全面的设计思维体系。

例如相应的课程设置：空间限定 → 空间分割 → 空间整合 → 空间延展。

一系列以"空间"为题的课程设计，通过宏观（城市）—微观（建筑）两个层面，使学生完成对建筑空间从"感知"到"认知"的转变。作业可以主要以模型表现为主，并要求附图面表现，表达方法不限。只有通过制作，并由图面表达出来，才能更透彻地理解空间，同时也训练了技法及构图。该训练旨在从素质教育的角度出发，根据建筑基础教学的特点，通过对空间的认知方法、造型方法、造型手段的研究和实践，探索学生如何最大限度地利用学习的技能，自我完善和发展。

2.3 设计媒介的更新——注重"实体空间建构"训练

"实体空间建构"不仅是对三维空间的感知过程，更是开发创造思维，培养设计表达和感知能力的基本手段。在国外，利用易操作的材料完成创造性思维，已经被纳入重要的职业训练过程，成为了建筑基础教学理论的一大支柱。各高校采取的主要措施有：

（1）加大三维空间模型的制作力度。新入学的学生虽然不一定都画过图，但大部分都做过模型，虽然大都不是建筑模型，但在做模型的过程中，对空间的体会与创造，实际上是大致相同的；因此，从三维模型入手，容易激发学生的兴趣与创造热情。

（2）学生进行建筑与空间的实体建构，并自行完成从设计到施工的全过程。设计阶段，除建筑设计外，还包括一定量的结构设计计算及实验、概预算等。在施工阶段，学生也可以参与建造以外的工期的计划、材料与工具的选购。

（3）建立（模型）加工车间。进行墙体片段、建构节点，以及大比例模型的制作。

3 小结

现代教育越来越深刻地认识到，观念、原则、方法的教育是建筑基础教育的核心，在建筑基础教育中建立起全面的"思维建构"观是非常必要的。因此，作为建筑的入门教育，应针对我国的教育现状，在学生的专业学习开始时，帮助他们突破习惯的思维定势，建立起全面的专业学习思维，全面培养学生捕捉问题的敏锐力、分析问题的洞察力和解决问题的创造力，并实现以感知力、观察力、分析力、理解力、想象力、表达力等六大能力为中心的思维建构新模式。

参考文献：

[1] 滕夙宏. 空间初体验——天津大学建筑初步课程中的建造教学实践. 新建筑, 2011（04）.

[2] 李伟. "思维建构"视角下的建筑设计基础教学. 2012全国建筑教育学术研讨会论文集. 北京：中国建筑工业出版社，2012.

The Exploration of Architectural Design Education
—Hanyang University, ERICA Campus, for Example[1]
建筑设计教育探索——以汉阳大学安山校区为例

Inha Jung
Professor, Department of Architecture, Hanyang University, Ansan, South Korea
汉阳大学，韩国

摘　要 本文旨在简要描述韩国汉阳大学安山校区的建筑设计教育的特点，及其在韩国认证体制下的转型。汉阳大学提供由KAAB（韩国建筑评审委员会）认证的五年制专业学位，教育目标为培养"能够引领全球化浪潮的创新性综合性专业人才"。教学计划保证毕业生有扎实的专业技能、批判的思维能力，并且有能力在快速变化的社会背景下辨明多样的职业道路。教学计划的核心部分为建筑设计的教育，由五个年级组成，使得学生可以循序渐进地学习建筑设计。为此，我们提供了10门设计课程及54学时学分，其中包括五年级的设计工作室阶段。常规的设计课程之外另有课外项目作补充，如跨年项目和国际建筑校友工作室。
关键词 建筑，设计，教育，汉阳大学

Abstract This essay aims to briefly describe the traits of architectural design education at Hanyang University, ERICA campus, and their transformation within the framework of Korea's accreditation system. Hanyang University is offering a five-year professional degree accredited from the KAAB, whose objective is to raise 'innovative and integrated professionals who can lead the globalization'. The program ensures that graduates are technically competent, critical thinkers, and capable of defining multiple career paths within a rapidly changing societal context. Architectural design education occupies the core part of the program. It consists of five levels which enable students to learn architectural design step by step. For this purpose, we are offering 10 design courses and 54 semester hours of credit, including a Level V design studio sequence. This regular design education is complemented through extracurricular programs such as Cross-Year Program and Architecture Alumni International Workshop.
Keywords Architectural, Design, Education, Hanyang University

1 Introduction

This essay aims to briefly describe the traits of architectural design education at Hanyang University, ERICA campus, and their transformation within the framework of Korea's accreditation system. Architectural department as one of three educational units at the time of foundation has run parallel with the university's history. Since its foundation in 1939, Hanyang University has made a remarkable achievement in all aspects of its teaching and research activities, and is striving for competitive status with internationally top-ranked universities. Hanyang has been committed to academic and professional excellence, with graduates playing active roles in areas of expertise, both domestically and internationally. The university has two campuses, at Seoul and at Ansan, in which 25,000 undergraduate and 8,000 graduate students including over 1,300 international students are currently enrolled.

The ERICA Campus at Ansan was opened in March, 1979. The campus is located in the central part of the West Coast of capital area which has been developing into a business hub of Northeast Asia. Despite its relatively short history, the ERICA Campus has successfully practiced development strategies and currently performs as a role model for other institutes in terms of an education reform. The ERICA campus also pushes

---

1　本文为"2011中英韩建筑教育国际论坛"宣读论文。

ahead a specialization policy pursuing an efficient combination of education, research, and industry to create synergy effects. Based on the network that covers universities, R&D centers, and companies, ERICA campus opens up new possibilities in merging new knowledge and technologies.

School of Architecture and Architectural Engineering (SAAE) was established as a department of Engineering College with 48 students in 1985. Afterward, the SAAE developed into two majors, architecture major and architectural engineering major, and the number of enrolling students increased into 73. Since the freshmen have to decide their majors after finishing their first year, the school provides them with diverse programs to find out their own aptitude. The education program of the two majors was organized to meet the requirement of the KAAB (Korea Architectural Accrediting Board) and the ABEEK(Accreditation Board of Engineering Education of Korea). The architecture major is offering a five-year professional degree which consists of five domains: general education, cultural context, design, technical systems, and practice. Through this program, the architecture major aims at nurturing innovative and integrated professionals who can lead the globalization. As a result of the development and performance of the specialized educational program, the SAAE ranked as one of top four architectural schools at the evaluation of the National Committee of College Education in 2000 and occupied the second place at the 2001 evaluation of architectural schools by the Joongang Daily.

The school has been pursuing diverse strategies of globalization: a Joint Lecture Program with the National University of Singapore has been successfully carried out since 2003. About 20 students of the two universities are swapped to spend one semester and get an opportunity for internship in Singapore and Seoul every year. The school has also hosted International Architectural Olympiad authorized by the UIA, in a bid to select talented high school students through design competition. Through this event, students can have the opportunity to experience university-level architectural program before they decide their majors. The school also have joint degree program with Illinois Institute of Technology, US, sending about 3-4 students to the university. This program help students to earn dual degree from the two universities, which can reduce the amount of time required to be spent at each.

2 Objective of Architectural Education

Hanyang University is offering a five-year professional degree accredited from the KAAB, whose objective is to raise 'innovative and integrated professionals who can lead the globalization'. The curriculum ensures that graduates are technically competent, critical thinkers, and capable of defining multiple career paths within a changing societal context. To accomplish this goal, we set forth four principles of architectural education:

1) Acquisition of a range of professional skills
Hanyang's architectural program aims to develop students' ability to communicate architectural ideas in foreign language, and to demonstrate architectural ideas in drawings or through appropriate media including photographs, models and information technology.

2) Understanding of architecture's historical, socio-cultural, and environmental context
Our program attempts to have students understand the interaction between traditional values and environmental factors that exists in individual or collective societal condition, and the theories of sustainability in making of architecture and urban design.

3) Architectural design including technical system integration, health and safety requirements
Our program promotes students' ability to integrate structural system, environmental control system, and construction materials and assemblies in integral building design.

4) Comprehending an architect's roles and responsibility in society

The architectural education program of Hanyang University prepares students to practice and assume updating roles in the context of increasing cultural diversity, variety of clients and regulatory issues, and expanding knowledge based on the profession.

3 Accreditation and Architectural Program

As already mentioned, Hanyang's architectural design program was accredited by the KAAB in 2009, which is the only accrediting board in Korea that is recognized by both the Canberra Accord and UNESCO-UIA. The KAAB is providing the backbone for architectural education in Korea. The establishment of the accrediting board resulted from the globalization of service markets in the 1990s. In particular, as the General Agreement on Trade in Services (GATS), a treaty of the World Trade Organization (WTO), entered into force in January 1995 as a result of the Uruguay Round negotiations, members agree to extend the multilateral trading system to service sector. This means the removal of barriers in architectural service markets.

In response to this change, Korean architectural associations made multifaceted efforts to meet international standards in architectural education. Prior to this time, Korean architectural education was framed by the intermingling of design and engineering education under the influence of Japanese system. The first step toward the educational reform was taken by changing the four year curriculum to five-year one and creating professional degree. The second step was to establish the Korea Architectural Accrediting Board (KAAB) in January 2005, whose mission is to set a guideline for the professional degree programs and to induce the fulfillment of the programs through accrediting and consulting. As a result, since the very first accrediting in 2006 until 2009, 34 out of the total of 74 5-year professional-degree programs in Korea were approved as candidates for accreditation. 23 out of the 34 candidates applied for accreditation, and 19 programs (professional master degree Programs included) were accredited.

The change of architectural program at Hanyang kept in pace with this tenet. The five-year curriculum was first introduced in 2002. After consistent efforts to meet high standards for accreditation and to obtain mutual recognition of their professional degrees internationally, our department succeeded in getting accreditation from the KAAB in 2009.

4 Architectural Design Education at Hanyang University

Architectural design education occupies the core part of the program. It consists of five levels which enable students to learn architectural design step by step. For this purpose, we are offering 10 design courses and 54 semester hours of credit, including a Level V design studio sequence. Each level must have a minimum of eight semester hours and a maximum of 10 semester hours. Design is defined as analysis, synthesis, judgment, and communication that architects use to understand, bring together, assess, and express the ideas that lead to a built project. The contents of each level are as shown below:

LEVEL I: Students begin to learn design process methodology, and development of communication skills; and design literacy.

LEVEL II: Students can design simple buildings with an introductory understanding of construction and structural systems.

LEVEL III: Students focus on the programming, site analysis, and design of complex building case studies with qualitative technical input. They also have the capability of making design decisions in altering existing designed environment by way of renovating, rebuilding and repairing.

LEVEL IV: Students are able to synthesize complex building and multi-building complexes within the urban context.

LEVEL V: Students reach the mastery of data collection, analysis, programming, planning, building design, and building systems.

5 Extracurricular Program

The regular course can be complemented through diverse extracurricular programs, one of which is Cross-Year Program. This is the program that every student takes part in to design a given simple theme in outdoor space at the beginning of every semester. Students can develop an effective decision-making system to compete with other teams and actively seek alternatives. The vantage of the program is to give a precious opportunity for a winning team to actually construct their ideas in a given site. To solve unpredicted technical problems, students form an executive committee to control construction process, and, if necessary, call for the consultation of experts. In the process students can learn to integrate design, construction and material etc., and promote communication, team work and leadership.

Together with the Cross-Year Program, our department is hosting the Architecture Alumni International Workshop during winter break by inviting alumni architects who are making a brilliant exploit in the US and Europe. Through a direct contact with graduates, this workshop is not only giving a precious opportunity for students to learn new information on design trends and researches abroad, but also motivating their challenging spirit.

Fig.1 2009 Cross-Year Program in Yeongheung Island

References:

[1] Standard of Accreditation of the Korea Architecture Accrediting Board.
[2] Education Guidelines of the National Council of Architectural Registration Boards.

基于教学体系下的住宅建筑创新理念研究[1]
Study of the Innovative Ideas for Residential Design Basic on the Education Structure

陆诗亮　Lu Shiliang
韩衍军　Han Yanjun
崔元元　Cui Yuanyuan
哈尔滨工业大学建筑学院
School of Architecture, HIT

摘　要　随着中国经济的发展，住宅消费市场化进程的加快，住宅设计无论外观还是内部格局，都呈现出新的设计趋势。在哈尔滨工业大学建筑学院教学体系当中，住宅设计作为一门本科生专业基础课，有意识地引导学生具有超前意识。本课程是学生从二年级开始经过一年多的理论、知识和技法的建筑学专业课教育后，第一次进行与现实社会紧密联系的设计实践。课程区别于之前的任何建筑设计课程，表现为制约限定性强、超前意识强、社会应用性突出，强制学生必须熟练掌握符合国家规范的设计，同时针对现实社会中的具体问题教会学生如何应对。
关键词　住宅建筑，创新理念，教学体系

Abstract　In recent years, with the economic development and the residential consumer marketization accelerating both the appearance and internal structure in housing design have present a new trend. The complexity and comprehensive of architecture decide that a certain scientific literacy, an ability of conducting scientific research with the use of scientific methods and creativity are the musts to complete the turning from a student to a professional architect. Traditional professional courses just base on the teaching ideology that focuses to design, ignoring research method. That the student looses the interest in autonomous search and the creativity cannot get sustainable development will be the long-term result. Based on the teaching system and course content of the School of Architecture of Harbin Institute of Technology, innovative concept of residential buildings in China is analyzed and studied in this article. The article emphasizes that, to adopt the students' forward-awareness in design is a critical circle to the development of building. And from the innovation of subject, function innovation, form innovation, environmental innovation and technological innovation in five areas of residential construction, innovative ideas were studied.
Keywords　Residential Building, Innovative Concept, Teaching System

1 超前意识

所谓超前意识，是指教学过程中，不仅需要深化和巩固学生对已经初步掌握的建筑的功能、流线、立面构图、体量造型、场地设计、空间组合规律以及建筑设计程序、方式方法等基本问题的认知，更重要的是要求学生学会用科学辩证的分析方法来研究及预测当今社会及未来社会住宅的家庭人口构成、生活模式以及区域环境、历史文化、地方特色，要求学生与时俱进，增进对现实社会及国情的认知，普遍了解当今我国住宅设计中的难点及国家亟待解决的居住中的主要问题，针对国家近年对住宅建筑陆续出台的鼓励或限制性政策和规范措施，熟悉掌握住宅多种多样的空间组合方式、建设模式及建筑技术特征，设计出满足不同居住对象、满足不同生活模式要求的住宅形式，并对住宅的空间构成模式作探索性的研究，力争推陈出新，同时对国家的强制性建筑规范做到初步了解及掌握。

2 创新理念

2.1 选题创新

课程内容注重与社会实践的结合，注重与专业前沿问题的结合。历年的设计选题全部为城市具体区域内的实际工程项目，具有极强的限定条件，避免了传统的"纸上谈兵"的设计教学模式。设计者可以通过实地勘测、参观体验、调查访谈等方式进行准备，从工程技术建造以及国

1　本文为"2011中英韩建筑教育国际论坛"宣读论文。

家现行规范和地方现行规范入手学习建筑，进而实现设计上的创新，使设计过程真正达到由感性认识向理性认识的过渡与转化。如在近三年的教学中及时引入"历史地段居住建筑改造设计"、"中低收入者居住建筑设计"等具有探索意义并能解决实际问题的设计选题（图1、图2），通过对我国建筑实践中许多建成作品缺乏建造深度的一种反思，着重讲解分析学科前沿最新设计与专业课训练的必然关联性，使学生在即将步入"高年级"前能尽快了解本学科前沿领域的最新信息，从而调动了学生主动投入专业课学习的积极性，也加强了学生的专业进取心和创业雄心。

2.2 功能创新

舒适便捷是人们对住宅的基本要求。社会及经济的发展，不断地给功能"舒适便捷"增加新的内涵，住宅课程设计要求在这方面适度超前，以满足人们更高标准的要求。其具体表现为：

户型多样化。室内布局注重实用，以户型方正、舒适、私密为前提。阳光、绿化平台、冥想空间、步入式更衣室等西方前卫理念被普遍引进普通住宅，传统的餐饮空间、厨房空间由于现代人生活方式的转变，其功能正逐步退化（图3）。

针对固定人群需求。针对不同的年龄段需求、收入阶层、工作模式、生活特征，表现出不同的应对方式。例如针对青年人设计的30~40m²小户型公寓式住宅成为设计选择的重点，有的设计在面积限定下已开始向立体分割方向发展，利用空间设计的不同高差分隔出不同的功能区域，大大提高了空间的利用效率（图4）。

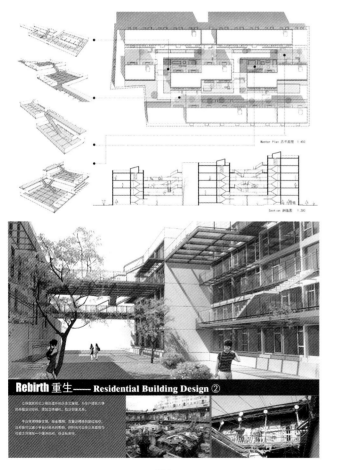

图1

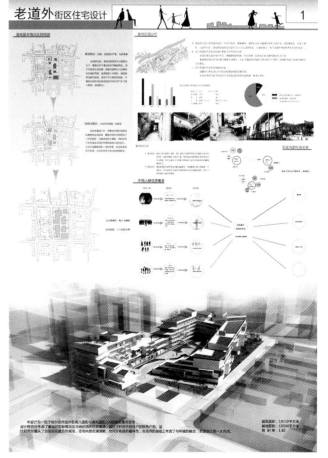

图2

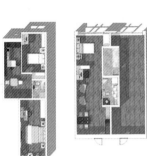

图3

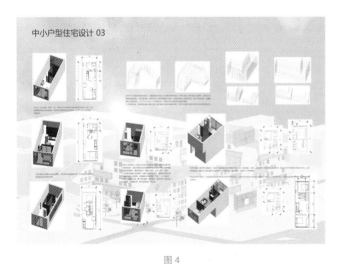

图4

入户交通方式多样化。对传统的单元式楼梯入户方式提出反思，取而代之的是集中式大堂、内外廊式入户、低密度电梯垂直交通入户、楼层阶梯式布局由室外屋顶花园直接入户以及由多种交通混合的入户方式。这些方式都反映出了设计者对居住趣味的关注，对浪漫生活的追求以及对人际关系的处理方式（图5）。

共享空间逐渐增多，住宅区除拥有集中绿化、园林、庭院、会所、架空层外，开始出现空中庭院，每隔几层设置一个共享空间。在开敞的空间内，植物、花卉相得益彰，使住户不出楼门即可直接感受室外的自然景观和邻里相亲的人情味。楼宇入口的大堂也开始向酒店大堂靠拢，专设洽谈、休息区，让住户拥有酒店式的享受。

2.3 形态创新

住宅形态设计是学生共同关注的重点，受时代的影响，近年，建筑形态设计表现出非线性化、参数化、装配式、高科技型等特征。

在住区规划形态方面，设计讲究整体性，大多建立集中绿地和相应的庭院设施，并且注意到了空间的完整性、趣味性和公用设施的齐全性。住宅郊区化成为新的热点，高层点式、高低搭配、多低层板楼和联排别墅（TOWNHOUSE）相混合，还包括一些其他类型的低密度住宅。

建筑形体关系变化多样，楼体则不仅有塔楼、多层，还有带电梯的板式小高层；户型面积出现两极分化，小户型和大户型都有较好的市场表现。

外部形象日趋简洁，外立面上已少见曾流行一时的繁琐的欧式符号，取而代之的是简洁明快的装饰色块及装饰构件。窗的开启扇面积愈见增大，大且低的窗台设计则拉近了人与自然的距离；全玻璃阳台代替了传统的水泥阳台，充满了时代气息（图6、图7）。一些设计注重想象空间的再创造：圆弧、镂空、转角、嫁接等手法的应用，使艺术与生活紧密结合。

2.4 环境创新

越来越多的学生认识到景观设计是住宅设计的新趋势。建筑与景

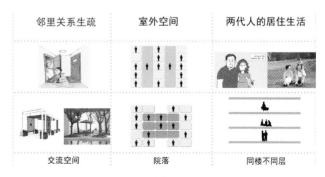

图5

图 6

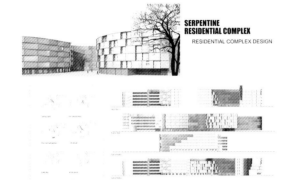

图 7

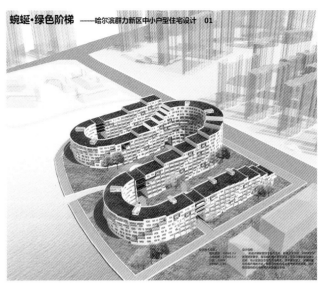

图 8

观之间有系统性和主题性标志，共同谱写景与境之间的内在联系。景观布局与建筑设计同步，景观环境与每幢建筑相互交融、渗透，丝丝入扣，相互对比、映衬，共同构成小区的风格。环境设计不能脱离建筑，只着重观赏功能而忽视使用功能，应结合建筑的使用功能，满足居民的休闲娱乐需求。同时，设计应让人进入景观内部，身临其境。学生近年来尤其重视屋顶平台的景观设计（图8）。

2.5 技术创新

新技术的应用。注重设计与结构、材料、设备、节能等其他专业的合理衔接，有效地培养学生的创新能力，逐步建立研究型大学设计学科"探索式"教学的新模式。

在结构方面，设计中除采用传统的钢筋混凝土结构外，轻型钢结构、张拉结构、预制装配技术、绿色生态技术纷纷出现，同时对膜材、轻型砌体材料、木质材料、金属材料等新材料的探索也跃然纸上。

另外，可动式住宅也成为当前学生研究的重点。当前的住宅设计在很大程度上对"发展"因素考虑不足，每户建筑面积偏小，而且一般都是定型设计，缺乏可改造性，越来越不能满足住户日益提高的居住需求。学生结合近年在建筑结构技术上的巨大进步，大胆构想了多种可变式住宅方案，实现了前所未有的空间灵活性，将同一面积标准的住宅设计为不同套型，以适应消费者不同的家庭结构，另一方面也可实现大套型分拆成小户型、小户型组合成大户型，满足不同住户的各种需要，大大增加了住宅作为商品的适应面（图9）。

此外，绿色生态技术也成为了应用的重点。一是太阳能在住宅中的应用，例如选用太阳能热水器、太阳能炉灶、太阳能辅助采暖等。二是风能在住宅中的应用，设想采用优化通风技术，使住宅系统的通风功能达到合理化、最优化，部分住宅还设想根据实际需要把风力用作动力，从而达到节能的目的。三是有目的地控制住宅噪声污染。在住宅内或住宅外建造一些噪声阻隔、噪声消除和噪声控制设施，以防止住宅随时可能遭受的噪声污染等（图10）。

3 结语

建筑大师勒·柯布西耶曾说过这样的名言："建筑是居住的机器"，然而我们目前的住宅建筑却远未达到机器般的精确和高效，这不能不说是建筑师们的遗憾。丘吉尔的名言:"人造住宅,住宅造人"道出了住宅的真正的价值所在。培养学生在设计中的超前意识,是对住宅发展以及建筑师培养至关重要的一步。建筑学科的复杂性、综合性决定了由学生转变为建筑师必须具有一定的科学素养,具有运用科学方法进行科学研究的能力,具有创造力。创造力的来源也是对事物的探求心,对事物细心观察、科学研究的产物。传统专业课程单纯以设计为主的教学理念,只注重知识内容的传授,不重视深入探求方法,长此以往,使学生丧失了对事物主动探求的兴趣,创造力也必定不会可持续地发展。因而在教学过程中,住宅设计教学理念强调应该关注和预测社会的未来,注重使学生在理论及设计实践上掌握设计的基本要领和设计规律,训练学生掌握一定的科学研究方法,培养其发现问题、分析问题、解决问题的能力,鼓励其批判和质疑的科学精神。

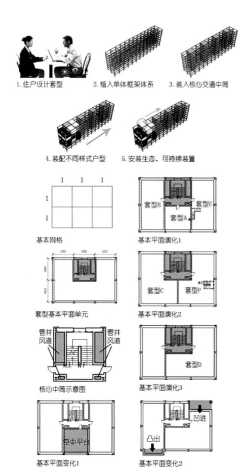

图9

1. 太阳能电池板

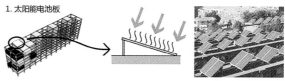

高效太阳能电池板能吸收太阳能,并将其转化为电能,供楼内日常生活的发电,起到环保节能的作用。

2. 雨水收集装置

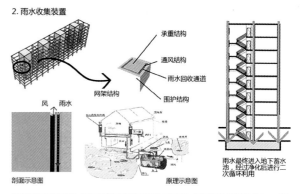

屋顶设置雨水回收装置,雨水会经过网架结构中的雨水回收通道,流入楼底的二次蓄水池中,一部分会用于小区内的绿化用水和其他辅助设施用水,另一部分经过净化处理后进入楼内二次循环利用。

3. 风道系统

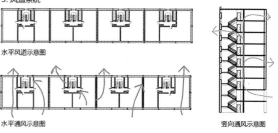

在全楼的网架结构框架体系中和楼层住户内墙之间都预设了相关的通风风道,风道的设置有助于室内在各个季节的通风,同时也起到了室内温度调节的作用,低碳环保。

4. 空中花园

每个楼单体都有一些网架中没有装配住宅,在这空间中引入室外绿色景观,从而具有良好的生态意义。

5. 智能化系统

小区通过智能化系统实现了用住户指纹即可开门、停车、购物、网上购物等多种活动。

图10

Architectural Design Studio Teaching in the Media Age
——Working with Uncertain Futures[1]

传媒时代的建筑设计工作室教学
——与不确定的未来一起工作

Renata Tyszczuk
School of Architecture, University of Sheffield, SheffieldS10 2TN, UK
谢菲尔德大学建筑学院，谢菲尔德 S10 2TN，英国

Abstract The age we live in is one that is conscious of our uncertain futures. The potential threat of rapid global environmental change has prompted a reconsideration of architectural practice – with questions of how practice should transform, how architects could reevaluate their commitment to society, how to think again about response and responsibility. Architectural design studio teaching offers an opportunity to rehearse the future: thinking through what might be possible in order to act in the present.
The dramatic acceleration of processes of urbanization and the radical instability of the environment demands that our entire approach to making and re-making cities be reconsidered. The prospect of an uncertain future in the current context is the major challenge for all those charged with designing, building and maintaining relatively enduring structures and communities on the earth's surface. One approach might be to think of design in terms of improvisation which might serve as a context for exploring possible futures.
The presentation will draw on examples of projects from the 'Architecture and Interdependence' Masters design studio at the School of Architecture University of Sheffield. The design studio was linked to the Interdependence Day research project and explored the challenges faced by future practitioners in the context of global environmental change. The studio environment aims to nurture an intellectual commitment to practicing the relations between diverse disciplines, different ideas, distant people and places. The projects and scenarios rehearse possible futures and explore the potential for an improvisatory practice.
Keywords Global Environmental Change, Pedagogy, Uncertainty, Improvisation

摘 要 我们生活的时代是一个对不确定的未来有意识的时代。全球环境迅速变化的潜在威胁促进了对建筑设计实践的再思考，诸如实践如何改革，建筑师如何重新评估他们对社会的承诺，以及如何再一次地思考回应与责任。建筑设计工作室教学提供了这样的一个机会来排演未来：思考为了应对未来，什么是可能的。城市化戏剧性的加速进程以及环境的不稳定性需要我们重新考虑如何建造城市。在当下语境下，不确定未来的前景是那些施加于建筑、维持相对持久的结构和社区的人的主要威胁。一种方法可能是考虑设计的即兴创作方面，可能会作为探索可能的未来的语境。
关键词 全球环境变化，教育教学，不确定性，即兴创作

Working with Uncertain Futures

What is it to live in an 'age' and therefore to have an epochal awareness? What is it to be conscious of one's place in history, and at the same time anxious about what the future holds? We seem compelled to designate the time we live in as an 'age' of one kind or another, whether media age, space age, or urban age. Indeed the twentieth century is often referred to as an 'age of anxiety' and our advent into the twenty first century has given rise to an 'age of uncertainty'[1]. We also ask questions of what the future holds and try to predict what the future might bring even though we know we can never really know the future. Architects are charged with planning, designing, and maintaining the spaces and institutions we live in, of providing shelter on the Earth's surface. This brings with

---

1 本文为"2011中英韩建筑教育国际论坛"宣读论文。

Fig. 1 whole earth

Fig. 2 Earthrise

it enormous challenges and responsibilities when working with uncertain futures in the context of global environmental change, because, architects are, in a sense, constructing for the unforeseen.

Space age

The space age has had significant political, scientific, technological and cultural impacts. The space missions have been synonymous with notions of progress and ideas of human ingenuity. However the remote images of our planet brought back to Earth by astronauts have also given impetus to renewed hopes of a collective endeavor, intergenerational equity and a shared future[2]. The Space age began with the launch of Sputnik 1 in 1957 by the Soviet Union and reached its apogee with the Apollo program. Just before Christmas 1968, as the Apollo 8 spacecraft completed the first manned circuit around the far side of the moon, the astronaut William Anders took a photograph of the Earth as it came into view. This photograph, dated 24 December 1968, NASA image AS08-14-2383 has come to be known as 'Earthrise'. 'Earthrise' had an immediate impact on the astronauts' return to Earth and quickly achieved iconic status[3]. The image was widely used in contemporary media, for example, as a backdrop on news and current affairs programmes and on the cover of the Whole Earth Catalog, marking its significance in galvanizing the emergent environmental movement[4].

Four years later on 7 December 1972 the astronauts on the Apollo 17 mission took a sequence of photographs of the Earth from space, one of which, NASA image AS17-148-22727 has become ubiquitous. Commonly referred to as the 'Blue Marble', this image is most familiar as a representation of the 'Whole Earth' and as such a signifier of 'our common world'. Themes of the earth's fragility, finiteness, limited resources and interconnectedness invoked by the NASA images were already in place in Buckminster Fuller's notion of a 'Spaceship Earth' that lacked only an intelligible operating manual[5]. In 1987, The Brundtland Report Our Common Future made explicit reference to these images of the planet: 'from space, we see a small and fragile ball dominated not by human activity and edifice but by a pattern of clouds, oceans, greenery, and soils, ' thus fostering a sense of the human predicament that has since become synonymous with the concept of sustainable development[6]. The image of the whole Earth from Space has since become all too familiar, routine and repetitive, but NASA continues to provide us with new images that instantly beguile as evidence of technological prowess and an ever-expanding universe and question our approaches to the future.

The Earthlights image, a composite satellite image of the Earth at night is an extraordinary map of the time we are living in Earthlights needs to be considered as dynamic image and one that reminds us of the interdependence of human

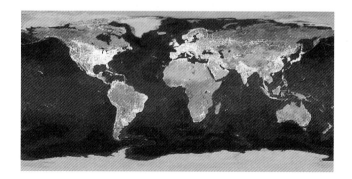
Fig. 3 Earthlights

and planetary activity. As the Earth moves across the sky it lights up at night; revealing concentrations of cities. But as some lights go on–as new cities grow and industries emerge in Brazil, India, China, others fade and disappear–those of the shrinking cities that mark industrial decline in northern Europe and North America. And we may also be aware, although still invisible on this map, that other lights flare dangerously on the Earth: those of recent disasters in 2011, the explosion of nuclear reactor in Fukushima or the forest fires in Australia. The shift in recognition between these three different renditions of our planetary home from space–Whole Earth, to Earthrise, and finally to Earthlights–reveals also a transformation of an imagination of an integrative planetary image of our common world to one that is dynamic and contingent, made by, and increasingly threatened by human action and uncertain conditions.

Geological age

The upheavals of our manufactured landscapes have prompted popular science writers to imagine a future world where cities remain as remnants in the strata like fossils of dinosaurs from a different age. In The Earth after Us, (2009), geologist and popular science writer Jan Zalasiewicz writes, "technological and natural processes have already become so inextricably interlinked that our actions now will literally be raising mountain belts higher, or lowering them, or setting off volcanoes (or stifling them), or triggering new biological diversity (or suppressing it) for many million years to come."[7]

In the future he imagines the discovery of a Human Event Stratum, partly destroyed by erosion and other geological processes, that lies buried between layers of sedimentary rock and varies from a thin sliver to several metres in thickness. It holds the fossilised remnants of cities, or 'urban traces', "[...] compressed outlines of concrete buildings, some still cemented hard, some now decalcified and crumbly: of softened brick structures: of irregular patches of iron oxides and sulphides representing former iron artefacts from automobiles to AK-47s: of darkened and opaque remnants of plastics: of white, devitrified fragments of glass jars and bottles."[8]

Mike Davis recently announced that, "Our world has ended"[9]. He was also referring to a new geological stratum but this time not a fictional one. The so-called 'whole earth' or Holocene era has been officially supplanted by the Anthropocene era, as confirmed by the Stratigraphy Commission of the Geological Society of London in 2008[10]. In addition to the build up of greenhouse gases, this new stratum is defined by human landscape transformation exceeding natural sediment production; by the acidification of oceans; by the relentless destruction of biota, and above all, by radical instability. The naming of the Anthropocene positions humans as the driving force of change and as such capable of epochal shifts, yet at the same time this also undermines all human constructions by reminding us that they are inherently fallible and unstable.

Age of predictions

In terms of discourse around climate change, we might also refer to the current period as an 'age of predictions'. IPCC reports have over the last decades predicted numerous possible future scenarios that may come about as a result of global environmental and anthropogenic, man-made climate change[11]. Climate change discourse concerns projections and predictions about the future of the Earth and climate scientists continue to offer us a range of plausible scenarios–most of them catastrophic. The work by the IPCC and UNFCC has been about the idea of imagining, and of predicting the future and over the last decade there has been growing confidence and force of argument in IPCC assessment reports about the future, including the speed and magnitude of climate change, and the role of human action.

Most discussion about climate change is framed in terms of trying to stop it happening, to avert the myriad catastrophes that threaten our age. This type of thinking produces little more than anxiety. But there is a more promising alternative, when thinking about working with uncertain futures, which is to think in terms of improvising[12]. Improvising suggests acting in response to the particular context or environment with what is at hand in more adaptable, spontaneous ways. We tend to think about our place in time along with dominant ideas of progress in fixed material terms, whereas if we rethought our histories in terms of the changeable and the makeshift, perhaps we would be better placed in dealing with the concerns of our age. It is worth bearing in mind that neither our constructions nor our histories are impervious to change but could instead be considered as "unaccomplished potentialities."[13] The to-and-fro movement of improvisation suggests looking back in time, as well as forward. When we consider human history we can find evidence of many resourceful responses to earthquakes, tsunamis, and other natural disasters[14]. Perhaps by shifting our attitude towards these events, and by appreciating that human existence on the planet is always provisional, resourceful and canny - that instability is in the nature of our existence - we might approach today's challenges differently.

Urban age

We are living at a time when urban populations have outstripped non-urban throughout the world, leading to many announcements of an 'urban age'.

The demographic pressures of rapid urbanisation and the growth of megacities coupled with environmental threats of rising carbon emissions and global poverty make the challenges that urban regions face difficult to comprehend and respond to. The dramatic acceleration of processes of urbanization and threats to the environment demand that our entire approach to making and re-making cities be reconsidered. The use and consumption of human and natural resources, conditions of work, human rights: all of these long-established human concerns need to be thought through in the light of the need to mitigate and adapt to future change. Technical adaptations and accounting systems are valuable and necessary, but have proven a far from sufficient response to the imaginative and intellectual challenges posed by global environmental change. The prospect of an uncertain future in the current context is the major challenge for all those charged with designing, building and maintaining enduring structures and communities on the earth's surface. Architects and urban design professionals have traditionally explored cities in ways which demand coherent solutions, solid constructions and certain states and yet cities and society are inherently unstable and unpredictable. Built environment professionals now find that the 'urban age' means working under conditions of increasing ecological, political and social uncertainty.

Uncertain Futures

One thing is certain: that whatever 'age' we are in, whether space age, media age, geological age, urban age or age of predictions, we are always dealing with uncertain futures. Uncertainty is here to stay. Architecture, it seems, in the midst of current fears about globalization, resource depletion economic crisis and environmental change, has never been more concerned with the prospect of an uncertain future and with questions of how practice should transform, how architects could reevaluate their commitment to society, how to think again about response and responsibility[15]. What seems to be important now is the way in which architecture might inhabit the dislocations and uncertainties of the present differently to think of them not as problems to be solved but as transformative realities that might lead to new ways of accommodating the future and building resilience to change. As Elisabeth Grosz writes: It is not [...] the reconstruction of the past that helps explain our present, but an understanding of our present, and its dislocations, that helps bring about unknowable futures[16]. Perhaps the 'media age' with its promise of global solidarity and new forms of technological connectivity might allow us to convene in hitherto unprecedented ways. While it already encourages multiple expressions of possible futures it might also foster shared enterprise along with shared uncertainties giving us scope to approach the design of the future in many different ways.

Rehearsing the future

Architectural design studio teaching is about rehearsing the future–a place to explore the shared uncertainties and visions concerning a future we cannot predict. Rehearsing the future is not about trying to imagine a wholly new future–by first creating some kind of blueprint and then determining its outcome–but is about understanding the present in the light of the possible. Above all it is about an engagement with the present scope of architectural practice and its potential for experimentation and transformation. My design studio teaching draws on a critical pedagogy that invites attention to entangled cultural, social, economic and ecological processes and a practice that considers the potential of provisional projects. It understands design as a strategic tool for working with unforeseen circumstances. The studio environment provides a place and framework to rethink the categories and assumptions made about architecture, and also about our interventions and responsibilities as architects.

As part of a collective and discursive approach, different attitudes, positions and interventions have been encouraged in the studio that in different ways have the potential to test and negotiate agency, social understanding and political action contributing to a design process that is transformative.

Mappings and Explorations by Studio Six

Fig. 4 studio six publication

The studio has provided a space for imaginative deliberations, challenging our preconceptions and transforming our understanding of the present. The process has been non-hierarchical: inviting, welcoming and provoking participation in the research of many different specialists and non-specialists, users, students citizens, and celebrating the diversity of approaches: "The projects embrace the essential instability, the sometimes bizarre and even threatening connectivity of the world. They are prepared to take a while for ideas to meld, for practices to form, for conflicts to emerge, for things not to work, or to work out in unexpected ways. They encourage thinking about how to produce work without predetermining its outcome and refuse to define space as programmable within the normative logic-dominated techniques of power."[17]

Improvisatory practice

My integrated research and teaching in the design studio has developed an approach to design I have termed improvisatory practice. This is a practice that 'rehearses the future'. It draws on both the pragmatist philosopher John Dewey's process of 'dramatic rehearsal'[18] by which dilemmas arising in situations are resolved, and an understanding of Aristotelian phronesis, or practical wisdom- which characterises the ability to reflect on and choose a mode of action in unforeseen situations. Architects need to develop their ability to improvise. Thinking through scenarios for future cities does not simply rely on future visions or re-framings but asks questions about the different skills required in the making of future cities, above all skills of responsibility adaptability and improvisation that are needed in the present: the ability to be able to work with what is at hand or available and at the same time think through the consequences of design.

An improvisatory practice relies on the ability to move to-and-fro between the past, present and future, between the everyday and possibilities and between more speculative data and transformative realities. Rehearsing the future does not mean that we attempt to forestall the future or create or

Fig. 5 Studio 6 mapping Nowa Huta

determine a new future according to our design, but instead that we are ready to respond to what the future brings. Ingenuity, spontaneity and resourcefulness are important in improvisatory practice but also a willingness to accept failure—to see design as a process of trial and error, of to-ing and fro-ing. If part of the responsibility of architects is the design and construction of habitable spaces for different users, respecting appropriate and sound use of materials, guided by needs of different situations, and in awareness of issues of sustainability, a recourse to an understanding of 'dramatic rehearsal' fosters the most important aspect of their responsibility, that of 'context and consequence'—the social context of their work, who they build for, why and the long term impact of their work[19].

Live Projects
There are many different ways to approach the issues of response and responsibility in design, some of which have been explored through Studio design projects and some through the Live projects programme at the School of Architecture in Sheffield[20]. Live Projects explore the skills and techniques needed by future practitioners through the idea of 'liveness' working with clients and communities outside of the academic institution. Through a careful consideration of the context they are also always concerned with future and continuity. Some Live Projects have become long-term projects which by initially nurturing small initiatives, others engage with existing practices in the city and region making them available to a wider public. One example, a commission by a local community group, was The Sheffield Food Network project, a website that was about developing a virtual meeting space that could connect all those interested in sustainable food practices in the city and region: growing, foraging, producing, cooking, selling[21].

The design studio projects do not necessarily always produce designs for buildings. They are as concerned with the unbuilt and the unbuildable—the social and cultural production of space—as they are with the material constructions of the built environment. These projects are characterized by their hybrid nature, both in terms of the questions posed—whether technical, ethical, political or ecological—and in terms of the range of people involved. Such projects inevitably work at all scales, in many media, and are produced in collaborative processes of design. From the smallest intervention to the largest-scale proposition—all projects are asked to consider the consequences and implications of design for others in the immediate term and in the long term, traversing near and distant futures.

Architecture and Interdependence
The design studio is the place where the traditional knowledge and skills of architects need to be developed to meet the challenges not only of our 'age' but also future ages. Consequently, in the design studios I have been involved with I have sought to open possibilities for design practice that are less determined and more improvisatory. The ambition has been to nurture a studio environment that considers interdependence: an intellectual and ethical commitment to practicing the relations between diverse disciplines, different ideas, distant people and places. It is an experimental and transformative environment that rehearses possible futures and explores the potential for change.

The design studio 'Architecture and Interdependence' (2006-2009) at Sheffield, was developed as part of the Interdependence Day research project, focused on taking a fresh tone and approach to issues around environment, development and globalization[22]. In this studio we explored the concept of interdependence in relation to the design of architecture, cities and the built environment. The studio included public and academic workshops, participatory art installations, video work and publications[23].

The dominant paradigm of responses to global changes has been of 'designing for sustainability'. The deployment of sustainability discourses in the field of architecture has tended to be limited to technical responses to specification or design problems on what is assumed to be a largely predictable planet. Reframing sustainability as interdependence involves an exploration of ways of working within planetary and human conditions understood as dynamic and provisional. It also suggests responses to global environmental change that provide an alternative to the prevailing culture of compensation, stabilising, or even valorisation of suffering that sustainability discourses tend to call up[24].

The research of the Interdependence Day project was informed by work in human geography that addresses the ethical and political implications of 'thinking space relationally'[25]. Doreen Massey's arguments in For Space, have suggested the possibility of thinking architecture as more provisional, less determined and more willing to work with the uncertainty of its outcomes. She asks '[...] that we recognise space as always under construction. Precisely because space on this reading is a product of relations between, relations which are necessarily embedded material practices which have to be carried out, it is always in the process of being made. It is never finished; never closed. Perhaps we could imagine space as a simultaneity of stories-so-far.'[26]

Thinking of architecture therefore, as provisional, as always under construction, and as a relation of stories allows for an exploration of the inevitable entanglements of humans and non-humans in designs for the built environment. Our studio design process recognized the importance of collaborative and improvisatory approaches that sought to research and accommodate a range of different interests and perspectives. Students deployed the what-if of scenario-making and storytelling as a design tool for rehearsing the future. Construction of scenarios enabled students to move between complex urban environments to intimate architectural details while seeking to understand the intertwined cultural economic and ecological processes. Considering ecological and social interdependence in design work has prompted a way of rethinking architecture and its categories, assumptions and systems. This is especially important in the current context where the social, economic and ecological, or more broadly, the ethical impacts of the development of cities are under question. These more provisional studio projects have been able to test larger themes that question, for example the economics, politics and ethics of transition[27] through a more immediate engagement with intentions, desires, site and audience. The projects ranged from collaborations with local community groups in Sheffield on urban agriculture through

Fig. 6 ID UKMap

to more theoretical research propositions on the re-use of local post-industrial sites to international concerns such as migration, knowledge economies and politics[28].

In 2009 the studio investigated the entire UK as an 'Open City– Interdependence UK'. With this project we were attempting to reframe the way that cities in the UK were presented by drawing on the many strands of thought about cities that had emerged from the Interdependence Day project, which focused on our mutual responsibilities and dependencies. By re-imagining the UK as one 'Open City', the scenario we were invoking was the constantly shifting, but actual and often unacknowledged condition of the UK. We tend to think of the UK as a collection of distinct cities arranged on an island. We have developed an 'island architecture' for survival, yet we remain dependent on global human and material resources, processes, services. Day to day life in the UK would be impossible without the human, cultural, economic and environmental contributions made by the rest of the world. We take all of this for granted, but with the challenges posed by environmental, financial and peak oil crises as well as the massive social and economic upheavals within cities caused by globalisation, it is clear that we can no longer remain oblivious to or evade responsibility for the people and resources we rely on. Thinking of the UK as an 'Interdependent' and 'Open City' begins to address this. By exploring the systems on which cities are totally dependent–transport infrastructures, distribution centres, oil and gas pipelines and above all people from all over the world–we can reveal that our island nation is more like an inverted diaspora, a complex web of dependencies.

The studio's contribution to the International Architecture Biennale Rotterdam in 2009 took these ideas further by presenting them in the form of a public information broadcast for the exhibition: 'Open City–Interdependence UK'[29]. The broadcast, aimed at a wider audience, attempted to expose how UK cities, far from being autonomous, exist within a wider ecosystem and included a range of scenarios of the future of the UK. Architecture and built environment professionals can no longer ignore that it is the systems and social strategies that need re-designing as much as the cultural assets, objects and urban landscapes that we tend to focus on. As Alex Steffen has argued: 'We are not only capable of understanding the systems around us, but of imagining and inventing their replacements, and mobilizing the constituency to make that happen.'[30] The credit, energy and climate crises demonstrate not only how human and global scales are inevitable intertwined, or how we think of the environment, but also provoke questions about how we encounter others. When energy and resources become scarce, how do we avoid hoarding but safeguard fair distribution? How do we avert civil unrest and geopolitical chaos?

The projects of the 'Architecture and Interdependence' studio were provisional responses: they were propositions which while consciously acting in the present nevertheless attempted to look to the responsibilities of architects for the future. But this wasn't a future that could be planned and determined by

Fig. 7 'Interdependence UK – Open City' 1–broadcast

Fig. 8 'Interdependence UK– Open City' broadcast

a few individuals but rather one that needed negotiating by many. The projects were also attempts to dissolve a way of thinking about architecture which subscribes to the myth of building certainties. The studio advocated architecture as an experimental and improvisatory practice capable of working with adaptive technologies as much as relational economies. In the words of Elisabeth Grosz, 'The radical role of the architect is best developed in architectural exploration and invention, in recognition of architecture's and knowledge's roles as experimental practices. Philosophy, architecture, science are not disciplines which produce answers or solutions, but fields which pose questions, and whose questions never yield the solutions they seek but which lead to the production of ever more inventive questions. Architecture, along with life itself, moves alongside of, is the ongoing process of negotiating, habitable spaces'[31].

Constructing for the Unforeseen:

Scenarios for Future Harbin

Some of these ways of working were explored in the three-day design workshop at HIT in September 2011, which investigated the future of Harbin in the context of global environmental change. We started the workshop with the assumption that, 'Architectural design is always about imagining the future' and we asked a series of questions: 'What is the future of Harbin?'; 'How could Harbin be different under certain unforeseen circumstances and with shared uncertainties?' 'What if...?' It was important to respond to the dynamic rather than static character of Harbin: as a territory within which the interdependencies of economy, ecology, culture and technology, were constantly in flux and not predetermined. The workshop deployed the power of scenario-making as a way of exploring the contingency of architectural design and the potential of multiple storylines and visions.

Initial research on Harbin by the different participants in the workshop was drawn together into an 'Encyclopedia of Harbin', a compendium of useful and non-useful knowledge, made with full awareness of the chaotic realities of any city: 'This encyclopaedia is not meant to be exhaustive, that is, it is not a case of stockpiling as much as possible but instead selecting and finding the unexpected, the interesting, the amusing, the bizarre and the essential, in some suggested categories.'[32] The interdependent categories suggested were: food, trade, waste, climate, topography, ownership, politics, ecology, culture, and technology. Each participant in the workshop worked on a different scenario for a 'Future Harbin' by first analyzing the existing conditions in the city and the region according to their chosen category, and identifying any issues and concerns. The next stage of the workshop involved casting possible future projections based on multiple visions for the city. This involved a playful exchange of drawings and information and discussion with all participants in the workshop so that by the end everyone had left their mark and made their contribution to everybody else's project, resulting in not only strange juxtapositions but important insights. The final stage of the scenario-making workshop worked on translating found, borrowed and appropriated conditions of the city into the multi-scale urban interventions and multilateral constellations of future scenarios for Harbin.

The design workshop at HIT was an experiment in negotiating spaces, building capabilities and thinking through the consequences for Harbin, by asking students to draw out the potential of the future city and its diverse inhabitants and stakeholders. Explorations in studio design teaching are about teaching future architects to be adaptable and resourceful because it is not just our constructions– political, social, material– that might need to be adaptable and resilient in the future. Architects will need to be adaptable and resilient also. It is important to nurture the possibility of an improvisatory practice because the only thing that is certain when working with uncertain futures is that in the future, things will be different.

Fig. 9 Future Scenarios Workshop Harbin

Fig. 10 Future Scenarios Harbin map

References:

[1] Zygmunt Bauman. Liquid Times: Living in an Age of Uncertainty. London: Polity Press, 2007.

[2] Shiela Jasanoff. Image and Imagination: the Formation of Global Environmental Consciousness. C. Miller and P. Edwards (eds.). Changing the Atmosphere: Expert Knowledge and Environmental Governance. Cambridge, MA: MIT Press, 2001.

[3] Bronislaw Szerszynski. Nature, Technology and the Sacred. Oxford: Blackwell, 2005.

[4] Vicky Goldberg. The Power of Photography: How Photographs Changed our Lives. New York: Abbeville Press, 1991; Simon Sadler. An Architecture of the Whole. Journal of Architectural Education. vol. 61 (4), 2008.

[5] Buckminster R. Fuller Operating Manual for Spaceship Earth. Carbondale: Southern Illinois University Press, 1969.

[6] The Brundtland Report. World Commission on Environment and Development (WCED). Our Common Future. Oxford: Oxford University Press, 1987.

[7] Jan Zalasiewicz. The Earth After Us: What Legacy will Humans Leave in the Rocks. Oxford: Oxford University Press, 2009. Zalasiewicz was one of the first geologists to adopt the new term "Anthropocene." The term was coined by Dutch atmospheric chemist and Nobel Prize winner Paul J. Crutzen in 2002 in the journal Nature to mean the "age of man."

[8] Ibid. p. 189.

[9] Davis Mike. (2008) Living on the Iceshelf: Humanity's Meltdown. 26 June 2008; http://www.tomdispatch.com/; accessed 3.7.2008.

[10] Stratigraphy Commission of the Geological Society of London (2008, a) 'Are we living in the Anthropocene?' GSA Today, 18(2), 4-8. Stratigraphy Commission (2008, b) 'The Anthropocene Epoch: today's context for governance and public policy'; http://www.geolsoc.org.uk/gsl/views/letters.

[11] The last IPPC report appeared in 2007. The next report is due this year, 2011. United Nations Intergovernmental Panel on Climate Change, Fourth Assessment Report, Climate Change 2007 (Geneva, 2007).

[12] Robert Butler, Eleanor Margolies, Joe Smith, Renata Tyszczuk. Culture and Climate Change: Recordings. Cambridge: Shed, 2011.

[13] Richard Kearney. On Paul Ricouer: The Owl of Minerva. Aldershot: Ashgate, 2004.

[14] Nigel Clark. Inhuman Nature: Sociable Life on a Dynamic Planet. London: Sage, 2011.

[15] Florian Kossak, Doina Petrescu, Tatjana Schneider, Renata Tyszczuk, Stephen Walker. Agency: Working with Uncertain Architectures. Oxford: Routledge, 2010.

[16] Elisabeth Grosz. The Nick of Time: Politics, Evolution and the Untimely. Duke University Press, 2004.

[17] Renata Tyszczuk. Open Field: Documentary Game, in Suzanne Ewing, Jérémie Michael McGowan, Chris Speed and Victoria Clare Bernie (eds.). Architecture and Field/Work, Critiques series. London: Routledge, 2010.

[18] John Dewey. The Collected Works. Jo Ann Boydston (eds.). SIU Press, Carbondale, 1974.

[19] Jane Collier. The Art of Moral Imagination: Ethics in the Practice of Architecture. Journal of Business Ethics. 2006. 66: 307–317.

[20] Live Projects are the pioneering educational initiative at the School of Architecture at the University of Sheffield that has been running since 2000. Live Projects are student-led projects happening in real time with real people in a real context. Architecture students work in teams with a range of clients including local community groups charities organizations and regional authorities. http://www.ssoa.group.shef.ac.uk/

[21] http://www.atlas-id.org/sheffield-food

[22] The Interdependence Day (ID) project has arisen out of an extended programme of action research on media and environmental change by Joe Smith, and the architectural design teaching, research and art practice of Renata Tyszczuk. Project partners in the ID project are The Open University (Geography), University of Sheffield (Architecture) and the new economics foundation (nef). The ID project has been supported by funding from ESRC/NERC. The Open Space Research Centre at the Open University has supported the development of the web and print publications: ATLAS. See Renata Tyszczuk, Joe Smith, Nigel Clark, Melissa Butcher. ATLAS: Geography. Architecture and Change in an Interdependent World. London: Black Dog Publishing, 2012.

[23] Renata Tyszczuk, Joe Smith. 'The Interdependence Day Project: Mediating Environmental Change'. The International Journal of the Arts in Society. Volume 3, 2009.

[24] Renata Tyszczuk. Architecture and Interdependence: Provisional responses to global environmental change. Ethics and The Built Environment conference. University of Nottingham, 2009; published in conference proceedings ISBN- 13 9780853582632.

[25] Doreen Massey. Geographies of Responsibility. Geografiska Annaler Series B: Human Geography. 2004 (1): 5-18.

[26] Doreen Massey. For Space. London: Sage, 2005.

[27] S. Spratt, A. Simms, E. Neitzert and J. Ryan-Collins J. The Great Transition. London, 2009.

[28] See for example, Jordan Jay Lloyd's project 'Urban Pantry' with the community group Grow Sheffield; http://www.atlas-id.org/urban-pantry; and also Julia Udall's ongoing work with Sharrow Forum and the campaign for Portland Works http://www.atlas-id.org/portland-works.

[29] http://www.atlas-id.org/interdependence-uk_open-city-broadcast

[30] Alex Steffen. Transition Towns or Bright Green Cities?. 2009. http://www.worldchanging.com/archives/010672.html; See also Alex Steffen (eds.). Worldchanging: A user's guide for the 21st century. New York: Abrams, 2006.

[31] Elizabeth Grosz. Chapter 15. The Time of Architecture. Bingaman, Sanders and Zorach (eds.). Embodied Utopias: Gender, Social Change and the Modern Metropolis. London and New York: Routledge, 2003. pp. 265–278; pp. 275–276.

[32] From the studio design brief for the workshop: Dr Renata Tyszczuk. Constructing for the Unforeseen: Scenarios for Future Harbin. HIT PRC, 2011.

基于建筑模型的教学思考 [1]
Teaching Consideration Based on Building Model

周立军　Zhou Lijun
于　戈　Yu Ge
薛滨夏　Xue Binxia
刘　滢　Liu Ying
哈尔滨工业大学建筑学院
School of Architecture, HIT

摘　要　通过比较建筑模型与其他媒介的特性，提出建筑模型具有引导入门、空间分析、培养和建立合作关系、感受与体验、辅助研究、分析建筑群体关系和成果展示等7大教育功能。在建筑教学中大量引入模型制作可以贯穿从基础阶段到职业准备阶段的建筑教育全过程。同时，提出应科学理性地看待建筑模型的教育功能，恰当使用，并引导学生清晰认识其性质和制作方法。
关键词　建筑模型，建筑教学，教育功能

Abstract　By comparing building model and characteristics of other media, Seven kind of teaching functions of building model was put forward that are guidance, spatial analysis, training and establishment of cooperative relationship, etc. Model-making is introduced in the whole process of architectural teaching from junior stages to the stages of occupation preparation. Meanwhile, the paper proposed scientific rational treating and the appropriate use of teaching functions of building model, and guided students to recognize clearly the characteristic and making method of building model.
Keywords　Building Model, Architectural Teaching, Teaching Functions

　　近年来，我国建筑教育蓬勃发展，建筑学和城市规划教学在质和量方面都有了很大进步。从教学手段上看，很多建筑院校大量增加了实体模型的制作。在计算机模型广泛使用的今天，为何要强调实体模型的制作，张永和认为："与平面媒介以及计算机模型的不同之处是实体模型的物质性。制作实体模型的过程中，材料的选择和运用、构造和制作程序的设计与筹划等，都是其他媒介所没有的。这些性质也使实体模型与同样是物质的建筑建立起一个相对'实'的关系，非物质的媒介自然做不到。"通过与其他媒介的比较研究，进而得出建筑实体模型（以下简称建筑模型）的概念。

1 建筑模型的概念

　　一般意义上的建筑模型，是指使用易于加工的材料依照建筑设计图纸或设计构想，按比例制成的样品。建筑模型介于平面媒介与实际空间之间，将两者有机地联系在一起。建筑模型有助于设计创作的推敲，可以直观地体现设计意图，弥补图纸在表现上的局限性。它既是设计者设计过程的一部分，同时也是设计的一种表现形式。

　　在方案设计阶段制作的建筑模型称为工作模型，制作应简略，以便加工和拆卸。在完成设计后，可以制作较精细的模型——展示模型（图1），供审定设计方案之用。展示模型不仅要表现建筑物的尺度、造型、色彩、质感和规划环境，还可以揭示重点建筑房间的内部空间、室内陈设和结构、构造等。

　　建筑模型制作，与手绘和计算机操作一样，都是建筑的设计工具与手段。作为设计工具，建筑模型可以用来研究和表现。根据设计阶段与需要，模型可以是抽象的也可以是具象的；可以是专门研究设计某一方面的，如空间或结构；也可以是综合性的，和平面媒介很相像。欧美和日本的建筑院校都有制作模型的传统，作为一项教学手段，其所具备的教育功能贯穿建筑教育的始终。

---

1 本文为"2011中英韩建筑教育国际论坛"宣读论文。

2 建筑模型的教育功能

建筑教育跨度极大，主要可以分为两个阶段：①基础阶段——认识与方法；②职业准备阶段——设计与研究。关于基础教育，近年来国内建筑院校教学改革强调过程教育、教学互动等，试图改变以往重图面、轻逻辑，重形式、轻理性的现状；而职业准备教育，主要体现在高年级课程设计与施工图实习上。部分国内建筑院校在这一阶段将教学重点从形式的生成转变为对形式的建造的关注。这些从低年级到高年级的教学改革，无论目标如何，本质都是教学手段的改革，其表现为大量引入建筑模型制作。建筑模型的教育功能愈发得以凸显，通过对国内外建筑院校以建筑模型制作为主要教学手段的课程进行分析比较，我们将其主要划分为7个方面。

2.1 引导入门

基础教育阶段对建筑师的职业生涯具有举足轻重的作用，设计入门的顺利与否直接决定了学生继续专业学习的积极性和主动性。通过无主题限定，允许学生自由发挥的模型制作这一体验式教学课程，可以激发学生的主观能动性，使学生一改以往的被动式学习方式为充满乐趣的主动求知，养成正确的专业学习方法和习惯，建立对于专业学习的热情和信心（图2）。

2.2 空间分析

建筑模型能够帮助学生有效地理解和操纵三维空间。一般的建筑模型容易引人更多地去关注建筑的外部造型而忽视内部空间，通过模型对建筑室内外空间的对位、连接、尺度等进行分析，是其教育功能的重要方面。这一教学手段之所以有效，关键在于知识的输入过程合乎事物的认知逻辑，即从直观的三维形象转换到抽象的二维图纸（图3）。

2.3 培养和建立合作关系

建筑师职业艰辛，建筑教育更应注重素质教育。设计师是不断面对竞争挑战的职业，要求具备积极健康、坚韧不拔的心理和善于沟通与协作的能力。以往的建筑教育中，专业知识或者被当成谋生的工具以体现个人价值，或者被当成某种"边界"，而傲慢地拒绝"陌生人"的进入。但是，怎样利用专业知识与其他设计师沟通协作，解决各种错综复杂的问题，却没有得到相应的关注。通过以设计小组为单位参与课程设计与模型制作，可以有效地培养和建立学生的合作意识与协作能力（图4）。

图1 建筑方案的生成和选择过程

图2 美国某院校学生入学后的第一个作业

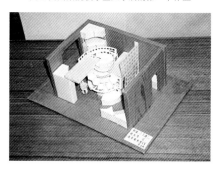

图3 哈尔滨工业大学学生"单一空间设计"作业模型

图4 中国美术学院的学生在建筑系馆门前制作1:1的等比例模型

2.4 感受与体验

对于教学来说，形式的生成和形式的建造是建筑设计教学的两大重要知识点。然而，目前我国的设计教学状况，从低年级到高年级，所有的课程设计都是在不同层面上重复训练着形式的生成，而形式的建造问题始终没有纳入设计教学中来。建造的知识牵涉到材料的质感、性能、重量、拼接方法、安装手段等问题，所以要让学生得到真实体会，实际建造训练是获取建造知识的最佳办法。例如瑞士苏黎世高等工业大学建筑系针对他们的建筑设计实践内容安排了"以砌筑为主的建筑形式建造训练"和"以钢结构为主的建筑造型训练"等课程。教师选择问题、设置建造训练的知识点与相应的途径，学生根据教师完成的图纸，在教师的实地指导下操作，模型和图纸一一对应，完成从真实材料和构造到二维图纸的转化过程。同时，学生也通过双手的操作体会到了材料与构造对造型的限定，引导了学生在工程技术的层面上对设计进行积极主动的思考，更重要的是培养了学生正确认识建筑设计的概念和内涵（图5）。

2.5 辅助研究

社会的发展对教育事业提出了更高的要求。城市建设需要大量具有宽阔视野和见识，知识面广，具有分析、判断和研究能力的有较高修养的建筑专业人才。这就要求学校的教学丰富多样，不只限于设计实践教学，也应包括设计研究教学。建筑模型是辅助研究的重要工具，在建筑和城市规划等设计研究方面具有十分宽广的适用范围。如图6所示，在限定同样大小和网格的底板上，学生通过不同的方案研究设计的多种可能性。

2.6 分析建筑群体关系

建筑教育的根本是将对空间和形体的学习、认识、理解和探索自始至终地贯彻到从基础阶段到职业准备阶段的教学思想和安排中去。同时应逐步由浅入深地学习和研究空间形体在不同尺度、不同时代，以及不同社会环境中的不同特性。建筑设计教学和建筑设计实践是紧密相连的。在设计实践中，经常需要处理单体建筑与周边既有建筑之间的关系，同一建筑师或不同建筑师设计的建筑群体之间的关系。建筑模型是进行此类设计分析与构思的最理想工具。图7为美国某院校学生在住宅设计中通过制作模型分析其与周边既有建筑的关系。

2.7 成果展示

建筑模型的最后一个教育功能是成果展示。作为建筑设计教学和实践、建筑技术教学和实践以及建筑历史教学等的成果，以建筑模型、构造与构件模型、历史建筑整体或局部复原模型的形式加以展示。这类展示模型具有较高的学术和研究价值，可以使学生获得直观的建筑认识，对建筑教育的发展起到了良性的促进作用（图8）。

3 结语

通过界定建筑模型的概念和特性并与其他媒介相比较，我们提出建筑模型在广义上具有引导入门、空间分析、培养和建立合作关系、

图5 日本长冈造型大学学生制作的模型室楼梯

图6 美国某院校学生制作的模型

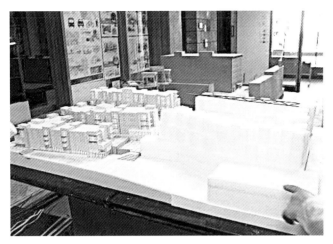

图7 美国某院校学生制作的住宅设计模型

图8 日本东京大学展示的密斯设计的德国馆内的柱子模型

感受与体验、辅助研究、分析建筑群体关系和成果展示等7大教育功能。在狭义上，建筑模型的教育功能主要体现为设计阶段的模型构思。模型构思可以从解决具体的空间构成任务，到解决任何一个抽象功能的形式组成问题。这些任务与教学大纲应相互交替进行，其目的是完整地培养学生的空间构成与组织能力，并使其以很高的创造性形式来完成并体现在他们的设计作品中。

在建筑教学中大量引入模型制作可以贯穿从基础教育阶段到职业准备阶段的建筑教育全过程。但同时我们也应科学理性地看待建筑模型的教育功能，恰当使用，并引导学生清晰地认识其性质与制作方法。

参考文献：

[1] 黄源. 建筑设计与模型制作：用模型推进设计的指导手册[M]. 北京：中国建筑工业出版社，2009.

[2] 项秉仁，丁沃沃，赵冰等. 建筑教育随笔[J]. 时代建筑，2001，增刊：30.

（图片来源：图2、图3、图6、图7周立军拍摄，图1、图4、图5、图8于戈拍摄）

Architectural Design Courses Teaching in Context of Media Era[1]
传媒时代背景下的建筑设计教学课程

Ivan Pazos
Assistant Professor, Hanyang University, Korea
汉阳大学助理教授，韩国

Abstract The lecture will evolve around the new desires for complex geometries as a result of new software packages. In particular it will be focused in the new developments of irregularity and asymmetry, in opposition to the classical notion of grids, order and symmetry. While many argue that this new geometrical order is just driven by the new computer programs and digital techniques, like a trend, others argue that the way humans perceive and create spaces is just about to change due to the new tools of thought available to us, the mathematical capabilities innate to mankind have been expanded by computing processes, almost like in a biological evolutionary process. Common algebraic regular geometries, clearly defined by equations and geometrical rules are shifting to non regular shapes and into bottom up self organizing systems. It almost feels like the Euclidean Orthogonal Space has finally evolved into a non Euclidean dream win which lines can never be parallel as they will eventually meet in the infinite. The teaching of digital architecture must deal not only with the learning of the new alphabets and computer tools, but also, and most importantly with the way we think and with the whole redefinition of the design processes and methodology itself.
Keywords Digital Design, Digital Architecture, Digital Teaching

摘　要 由于新的软件的产生，这个讲座将围绕着对于复杂的几何形状的向往来展开。它尤其将集中于与传统的网格、秩序和对称的古典概念相反的不规则和不对称的形体的新发展。然而许多人认为，仅仅是新的计算机程序和数字技术带动了这个新的几何秩序，就像一个趋势，而其他人则认为，人类感知和创造空间的方式就是改变。由于我们拥有新的思考工具，人类与生俱来的数学能力已扩大到了软件进程，就像是生物的进化过程。被方程和几何规则明确定义的规则的几何体，已经开始转变为不规则的形状和由下而上的自组织系统。这就好像是欧几里得正交空间最终演变成一个非欧几里得梦想，即两条线永远不是平行的，它们会在无限远处相遇。数字建筑的教学，不仅要学习新的基础知识和计算机工具，最重要的是学习思考问题的方法和对整个的设计过程和方法本身的重新定义。
关键词 数码设计，数字建筑，数字教学

We have recently witnessed new desires for complex geometries as a result of new software packages. Since computers have become an integral part of architectural design the range of tools available to designers has extremely increased, and with them methodological processes that used to be extremely complex have become available and sometimes very simple.

But not only new software packages, also hardware such tables, smart phones, hard drives, flashdrives, laptops, netbooks and wireless routers allow great mobility and new nomadic work styles. Also the web has changed the way we work, shop, socialize, communicate, access content, research and particularly the way we think, design and build. The computers have changed design in many ways, team work structures, management practices, construction processes, design techniques, access to information, computer graphics, algorithms, sustainability, engineering, scripting, parametric modeling, BIM and so on. This essay will however focus on the direct consequences of the use of computer and its links with the birth of new complex irregular curved geometries.

1　本文为"2011中英韩建筑教育国际论坛"宣读论文。

Architects have always been able to produce curved surfaces, but the tools they had at each moment in history defined the curvature type predominant for each historical period. In these days Software packages such as 3D Max, Rhinoceros, Maya, Revit, Grasshopper and many others allow the possibility of dealing with complex irregular geometries unthinkable before. On the other hand also software packages such Adobe, AutoCad, Archicad or Corel have totally changed the way we represent and the way we express ourselves, almost like if were to using a new code. We have not reinvented the language, but we are using new alphabets that unintentionally are influencing and evolving the way we speak.

Some new types of mathematical entities such as NURBS and Polygon Meshes have been developed in recent years by Software engineers. NURBS stands for Non Uniform Rational B-Spline, the complex name already implies that is a new type geometrical entity based on non uniform parameters.

Polygon Meshes are also known as Unstructured Grids, the name again defines what this type of tool generates, which are grids without a defined structure than can create the most impossible geometries. The software programmers created tools that allowed the creation and easy manipulation with extreme accuracy of non uniform unstructured shapes, unintentionally laying ground for new architectures that soon will follow.

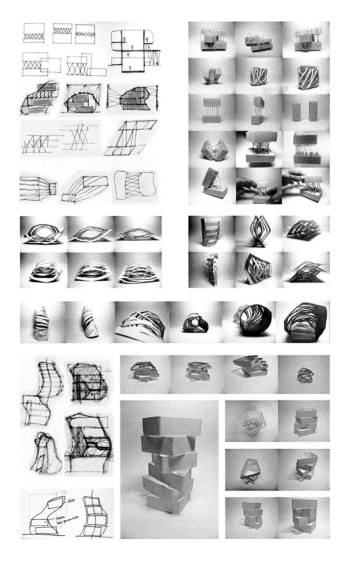

In particular we are witnessing new geometrical developments based on irregularity and asymmetry, in opposition to the classical notion of regular grids, order and symmetry. While many argue that this new geometrical order is just driven by the new computer programs and digital techniques, like a trend, I argue that the way humans perceive and create spaces is changing due to the new tools of thought available to us, the mathematical capabilities innate to mankind have been expanded by computing processes, almost like in a biological evolutionary process. Common algebraic regular geometries, clearly defined by equations and geometrical rules are shifting to non regular shapes and into bottom up self organizing systems. It almost feels like the Euclidean Orthogonal Space has finally evolved into a non Euclidean dream on which lines can never be parallel as they will eventually meet in the infinite. The teaching of digital architecture must deal not only with the learning of the new alphabets and computer tools, but also, and most importantly with the way we think and with the whole redefinition of the design processes and methodology itself.

The educational methodology of architecture has to change accordingly, by incorporating all this factors not only into the teaching process, but also into the design process. Students should be trained to use the computer not only as a tool of production but as a design tool. This process of thought is radically different for the traditional 2 Dimensional and orthogonal driven Euclidean projections Plan, Section and Elevation. In fact Euclidean methods of representation based on orthogonal projections have become obsolete for certain architectures, on which walls are tilted defying gravity, floor plates sloped and envelopes are curved with no differentiation between roofs and facades, and partitions and ceilings merge into a single entity.

However we should be aware of the risks associated with the use of the new tools. They are powerful and they are not a substitute to the architects mind. Architecture is based in creating spaces. Digital tools have their own dangers, and can get easily out of control, in particular when designers are not very familiar with the software and the computer design methodologies and processes. Architects create spaces, based on their clients needs, technology and spatial desires, the digitally dream have expanded our possibilities, but that doesn't mean that we will be able to create better architecture by using the computer and the web. Great architecture will always be great architecture, no matter how is produced, whether it has been created digitally, or crafted manually. Only time can dictate which architectures will stand time and become eternal, and whether or not they have been created digitally will be irrelevant. Let's use all the available tools to our best advantage. We should never forget that light, proportion and space are the prima matter of architecture.

建筑教育国际化培养模式的理论与实践探索 [1]
Internationalization of Architecture Education Training Mode on Both Theory and Practice

张姗姗　Zhang Shanshan
白小鹏　Bai Xiaopeng
白晓霞　Bai Xiaoxia
哈尔滨工业大学建筑学院
School of Architecture, HIT

摘 要 本文在分析我国建筑教育国际合作模式现状的基础上，首先从多元化的教学模式、国际化的参照平台、精细化的培养体系和复合型的成果评价体系四方面进行理论探索。其次，结合近两年的国际合作教学实践从以真实项目为载体、建立可持续的合作方式、从容的设计周期以及综合性的成果评价四方面进行了实践探索的总结与提升，并提出了可行持久的建筑设计国际化联合教学模式。
关键词 建筑教育，国际化，培养模式

Abstract This article first makes an analysis on the present conditions of the international education cooperation and then goes on to do theory research, including diversified teaching mode, International reference platform, Intensive cultivation system and composite type of achievement evaluation system. The third part of this article is practice research with the experience of recent two years, including real project practice, sustainable cooperation, guaranteed period for design and comprehensive achievement evaluation.
Keywords Architecture Education, Internationalization, Teaching Mode

1 引言

改革开放30年来我国的建筑教育得到了前所未有的发展，各建筑学院校也已经意识到我国的建筑教育发展和改革需要借鉴发达国家的先进理念和实践经验，吸纳国际一流的教学模式来促进我们自身的发展。建筑教育的国际化近些年已经取得了一些国际合作的成果，与此同时，也暴露出了许多问题，这也说明对于如何具体实践国际化，我们仍然处于探索的阶段。建筑设计是注重方法的教学，创新的教学方法是培养创新设计人才的前提。与国际知名院校联合引进一流的教学人才，可以开阔视野、学习先进的教学理念和教学方法，为培养国际一流的建筑学人才奠定基础办学条件。国际化教学应当是双方及多方共同输出、共同引进的过程，如果对国外模式和理念单向全盘盲目地接受，则缺乏针对性，甚至可能冲乱了自身文化的传承。在引进国外教育理念、设计理念的同时，输出与展示我国自身的特征，从而互相碰撞，共同提高，只有共同获益的方式才是稳定长久的发展模式。

2 国际化培养模式的理论探索

建筑学这一专业领域极其强调培养设计者的思维深度和眼界广度，而时代的发展又提出了培养建筑师的国际化视野这一更具有挑战性的目标，对于研究生培养，我们希望他们的国际化视野不要停留在对于信息资料的收集和整理上，通过体验式的学习过程、参与性的实地考察、面对面的多国交流，获得其设计的经验和灵感，往往能更好地形成一个建筑师的国际化思路和创作观。

2.1 多元化的教学模式

多元化的教学模式不仅是教学形式的多样化，更重要的是设计

1　本文发表于：2012全国建筑教育学术研讨会论文集. 北京：中国建筑工业出版社，2012。

思维的多元化与设计成果的开放，在教学中强调面向世界的建筑学术思想。在面对于同一题目，从不同的视角、运用不同的方法进行创作的过程中，不同国家的思维方式和解题思路产生了建筑创作的不同思维模式，产生了"混血"效应。在一次完整的项目训练中，如果能够使得来自不同文化背景、不同社会立场的参与群体充分展示自身的独特性，那么这些独特的观点、特征、手法集体碰撞之后，在彼此取长补短的过程中会形成全新的更为完善的模式。多元化的教学模式关注的重点在于培养学生自主发现问题、解决问题的能力，强调面向未来的思想和方法的交流，而非单纯的知识叠合，因此多元化的教学模式应当具有更强的自由性和包容性，允许师生有自主发挥的灵活空间，设计成果多样开放，力争在百花齐放中优中选优。

2.2 国际化的参照平台

国际化的参照平台既是学习和了解他人优点的平台，同时也是展示自身的机会。各参与国家及院校能够在该平台上紧紧把握建筑学科的最新发展动态，从而进行准确的自我定位。国际联合教学是以一个共同的项目、一个共同的标准进行完整的设计，形成各国之间、校际之间、多文化之间互为参照、互相促进的整体的教学结构，只有在国际化的参照标准中相关的师生才能够相对准确地定位自己在国际化的进程中处于什么样的位置。而对于最终成果的鉴定，必须有国际的参照系，否则参与者难以在国际化的平台上明确自身的优势与劣势，从而也就难以判断下一步的培养应朝哪个方向努力，如此，国际合作将无法发挥其最大的合作效益，甚至可能流于形式。

2.3 精细化的培养体系

国际合作在取得成果的基础上应向更加深入、更加精细的方向发展。建筑学教育当中，应针对不同的学生群体提出不同的培养模式。例如针对博士研究生、硕士研究生、本科生，在国际合作培养模式上应当有所区别，是倾向于全球视野的科研探索，还是注重设计思维的激活，在培养中应当明确目标，将培养体系精细化，针对不同的群体制定最适宜的培养模式，使得在有限的时间、有限的精力下学生得到最为高效的提升。

2.4 复合型的成果评价体系

复合型的评价模式指的是将外界评价体系与自我评价体系进行结合，从而更加全面地评价训练成果。在以往的设计训练当中，多以外界评价体系为主，尤其是多以单一群体的评价模式为主，单一群体背后的知识体系、社会立场等也往往趋于单一，如此，对于训练成果的评价视角难免有些偏颇。例如单一群体对于学生作业的评价仅代表了该群体的知识体系及立场评价，缺乏其他社会立场的评价，如业主、政府等。更重要的是，缺乏设计者彼此之间的自评体系。自评体系的缺失使得设计者彼此之间的发言权降到了最低，如图1所示。

3 国际化培养模式的实践探索

在国际化培养模式理论研究的基础上，哈尔滨工业大学建筑学院公共建筑与环境研究所近两年针对研究生教学进行了一定的实践探索。自2010年以来，加入了来自不同国家十多所院校共同组成的国际工作小组，该组织是以全球建筑共同进步为理念的教育机构。

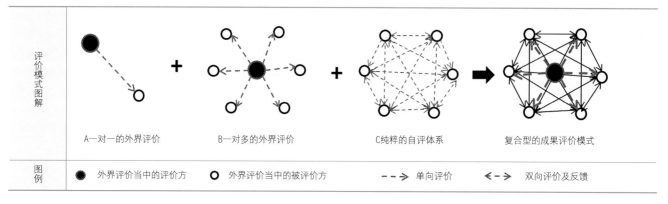

图1 复合型成果评价体系图解

3.1 以真实项目为载体

研究生阶段建筑设计的培养已经不简单是对于某一假定题目的求解，更重要的是对于真实问题的研究与解决。真实项目其设计背景所需要的思维理念及选取的技术具有极其明确的指向性，这样的项目具有一定的研究性质，通过该训练使得参与者体会并掌握如何通过研究解决陌生环境当中的实际问题。在训练中选用国外的真实项目为题，能够准确地体验国际工程的设计实况，从参与前期策划、自主地发现问题、完善任务书到城市设计、建筑单体设计乃至最终的技术设计，在这完整的训练过程中，参与者的思维深度、思维广度均可得到明确的提升，国外真实项目的训练为参与者走向国际设计搭建了模拟的平台。既有本地熟悉的思维，又有从外部世界观察当地的全新角度，两种不同的思维方式的碰撞，可以更为客观和全面地观察真实的问题。此外，与理想化的假设题目相比，真实的题目更可以调动学生的积极性。

2011年国际工作小组的设计研究为"法国欧什废弃兵营的复兴改造"，由法国拉维莱特大学选择，并经其他所有参与院校的认可最终确立。该项目为法国欧什废弃兵营的真实的复兴改造计划，因此得到了当地政府的大力支持。该项目包括城市设计、单体设计两个层面，同时涉及古城保护、建筑改造、文化传承等问题，属于综合性极强的题目，作为研究生国际化合作的真实项目具有重要的意义（图2）。

3.2 可持续的合作方式

建立稳定、长期的合作模式能对我国的建筑教育体系起到根本性的指导，而突发性的、闪烁的合作方式难以形成持续的、可传承的影响，在逐年的重复合作中可以寻找到恰当的合作方式，因此选择适宜的合作机构是建立稳定长期合作的前提。其次，彼此合作的动力源自对于共同发展和进步的追求，我们在看世界其他各国，同理他们也在看我们的发展，各国之间以期通过取长补短、资源共享等方式来促进彼此的进步，因此只有使得参与的各方实现多赢才可能建立可持续的合作模式。再次，合作方式便于教学计划的展开与实施才能在客观上保障合作的稳定，即应当寻求稳定存在的合作机构。此外，合作模式应当具有一定的开放性，不断吸纳新力量加入，使这种共同合作的组织能够不断壮大。

国际工作小组当中的各个国家的院校在这个平台上共同付出，共同收获，各国轮流作为东道主提供设计实践题目，这样对于每个院校的师生来说都是将"走出去"与"请进来"双向模式进行综合，群体合作，共同获益。这种合作方式的参与性更强，文化的多元性特征明显，属于多对多的方式，相比一对一或者一对多的方式更能碰撞出思想的火花。此外，出题院校可借助这样的机遇寻找到解决该国城市问题、建筑问题的新思路和新视野。

3.3 从容的设计周期

适当的设计周期为参与的师生提供深入思考和研究的时间保障，大多数的国际合作设计以短期、"快速"为主，在以往的训练中，时间上的差异是导致我们的设计无法深入的直接原因之一。我们主张在训练当中模拟国际的设计周期，同时结合我国的现实情况，将训练项目的设计周期定为介于二者之间，建议取6个月至1年的设计周期为佳。国际工作小组一般以6至9个月为设计周期。初期组织各参与院校师生进行实地调研，进行当年的首次交流，对该项目进行思想的交流。对于一个陌生的城市进行设计，首先应当尽可能地去了解当地的文化，从而做出更适合当地的设计，否则将演变成将本国、本校的理念强加于这个特定的基地。而后近6个月的时间进行各团队的独立创作，在实际了解题目之后，拥有相对较长的时间去进行深入的思考和反复的论证，这样更接近于真实的设计。从容的设计周期是设计深入程度的保障，可以让学生进行更为完整的设计，从规划到单体以及最终的节点设计，是整套的成体系的训练，而非简单地停留在灵感和创意的层面。此后再次齐聚进行设计成果的汇报与点评。

3.4 综合的成果评价

国际工作小组2011年作品的点评采用了自评体系（即参与的所有院校师生自身）与外评体系（即欧什政府官员代表）相结合的方式，这样使得评价的结果更为全面，使得大家从其他参与者身上发现解题的其他思路，并且能够有效地获取第三方的评价意见，从而对自身在国际平台当中所处的位置有更准确的定位。设计成果的点评具有三重意义：首先，成果的交流是整个国际工作小组当年训练成果的检验，即教育意义；其次，设计成果得到了项目当地政府的认可，对于解决该基地问题提供了宝贵的参考，即真实意义；再次，优秀设计对于此类项目形成了一定的示范意义。图2所示为国际工作小组的合作模式及其发展示意。

4 结语

在国际工作小组的实践当中，相关院校彼此学习，相互促进。这种多对多的国际合作模式为国际化教学提供了新的参考。在今后的设计教学当中，我们仍将坚持从理论和实践上探索国际教学模式

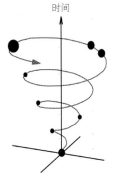

图解可持续发展的合作模式

- 总结经验 —— 不断完善的新一轮的合作
- 复合评价 —— 外评与自评相结合
  - 答辩与交流
  - 方案自述
- 从容设计 —— 2011年3月~9月
  - 深度思考
  - 多方案比较
- 多元交流 —— 调研与交流相伴进行
- 现场调研 —— 2011年3月赴欧什调研
- 真实项目 —— 法国欧什兵营改造
- 国际平台 —— 以国际工作小组为平台

A校方案

B校方案

C校方案

成果互评与欧什政府外评结合

调研现场

图2 国际工作小组合作模式发展示意及2011年合作程序

促进交流各方共同前进的方式，以期使国际化教学的平台更广阔、更完善。

参考文献：

[1] 朱文一，刘健，张晓红. 立足中国特色，培养建筑帅才，跻身世界一流——清华大学建筑学院面向新世纪的国际合作与交流. 从这里走向世界——清华大学国际合作与交流论文集. 北京：清华大学出版社，2010:16~23.

[2] 王维敏. 合作创新 建筑教育实现国际化. 中华建筑报，2011.

开放式建筑教育的深化拓展 [1]
The Deepening of Open Architecture Education System

张姗姗　Zhang Shanshan
蒋伊琳　Jiang Yilin
白小鹏　Bai Xiaopeng
哈尔滨工业大学建筑学院
School of Architecture, HIT

摘　要　本文对目前国内开放式建筑教育的现状进行了研究，总结出其模式化、表面化和片面化的一些局限，并从思维上对开放式教育的理论主体进行深化拓展，分析了开放式建筑教育表面和深层两个内涵，并通过对建筑教育的主体转化和建筑教育的思维拓展等一系列方法的探析，实现深层的开放式建筑教育，从而实现多元思维这一开放式建筑教育的目标
关键词　开放式教育，建筑教育，多元思维

Abstract　This article mainly talk about the open architectural education system, summed up some of the limitations such as fixed pattern, superficial and one-side mind. Make a deep thinking and expansion of the main body of open educational theory; find its surface meaning and the deeper meaning. Then realize the deep meaning of the open architectural education through a series of methods of transformation of the principle part in architectural education and expand thinking mode, and finally achieve the education goals of the open architecture of multiple thinking.
Keywords　Open Education, Architectural Education, Multiple Thinking

1 走向国际化的开放式建筑教育现状和局限

1.1 模式化的开放式建筑教育

中国的开放式建筑教育实践始于从1980年开始的清华大学的一系列国外联合设计课程，后来又发展出国际论坛、国际展会、国际联合竞赛和学生联合培养等一系列形式，到2000年栗德祥教授在新建筑上发表《呼唤开放式建筑教育体制》为止，中国建筑教育的开放性已经从最初的探索发展成具有一定规模和模式化的建筑教育组成部分。开放式建筑教育的具体实现途径也从最初非体系的外围探索，走向了正规化和制度化。

在其发展过程中，开放式建筑教育的模式也逐步形成，其核心内容无外乎创造多元文化环境，并在一个共同的载体上寻找思维的碰撞，以期达到一种思路的拓展、方法的提升和多元评价，其中的载体主要有虚拟的设计课题、真实的项目、国际热点理论问题、具体的技术措施等几个类型。

模式化的开放式建筑教育给我们带来了一个与国际接轨的途径，并通过一系列可靠的方法使得这个过程有了具体的操作和评价的可能，使得建筑教育的国际化这个概念得到了具体的实现，这一点毋庸置疑。但是从《呼唤开放式建筑教育体制》一文中我们也可以看到一些对于开放式建筑教育体制因为模式化而变得片面的忧虑，文中提出了开放的学科方向、开放的培养目标、开放的教学计划和开放的学术环境等一系列方法的外延概念，这些虽然还难以在当今的教育体制中实现，却展现出了"开放"这一概念的真正内涵。

1.2 表面化的开放式建筑教育

开放式建筑教育的实践历经了30年的过程，目前已经在全国的建筑院校中得到了广泛的关注和应用，但是在这其中也有一些不得要领、表面化的现象产生，必须引起我们的重视。

1　本文发表于：2012全国建筑教育学术研讨会论文集. 北京：中国建筑工业出版社，2012。

到底什么是"开放",是不是我们把门打开,请来几个外国建筑师或者工程师就开放了呢?答案无疑是否定的,开放的究竟是体制本身还是我们的培养客体?如果是前者,那么很容易实现,我们通过以上介绍的一系列载体引入多元文化就可以,但是这样的开放只是表面的开放。学生的参与度到底如何?他们学会了什么?他们真的形成了开放的设计视野和设计思路了么?结果往往不尽如人意。在我们引入这种多元文化的同时,如果不对课程设计进行跟踪,不对教学方法和结果进行评价,那么很有可能适得其反,学生们原本不成熟的设计思维没有得到发展,却被一个强势的设计思维取代,成了个人风格的牺牲品,只会人云亦云,何谈开放思维。就这一点来说,我们实践的很大一部分都流于表面,对于开放设计思维之培养这一真正的目标功效甚微。

1.3 片面化的开放式建筑教育

片面化体现在对于开放式教育理念本质的认识上,开放式教育是一种教育思想,而不是一种教育的外在表现形式,多元文化主体只是实现其思想的具体方法中的一个组成部分,但并不是这个方法的核心。能否形成一个开放式的学生培养模式,关键在于一个教育主体的常规力量的培养思路,这在很大程度上与一个院校老、中、青三代教师对于设计教学和培养目标的认识相关,他们是整个教学的主体,其他外请的教师和兼职的教师只是这个系统的辅助。要从这个层面落实开放式建筑教育思想难度更大,却能由内而外改变现在的教育现状,而不是借由外力给建筑设计教育披上一件开放式的外衣。对于开放式建筑教育思想的本质挖掘,也能给其带来新的活力,从而减少教育资源地理优势集中和多元文化交流的时间短暂不能持续等发展限制,跨越多元文化这一开放式教育理念的潜在限制因素,挖掘出开放式教育的本质。

2 由表及里——开放式建筑教育理论的新拓展

2.1 回归探索——建筑教育的主体转换

开放式教学应该让学生在一个开放的状态中重新找回好奇心,回归到探索这一充满无限可能性的学习模式中,有一位建筑老师曾在访谈中对他的教学方式作过一些描述,他带学生到苏州园林去参观的时候只是坐在一边喝茶,然后对他们说:"去找你们觉得有意思的东西,然后向我汇报。"这就是一种开放式的探讨问题的方式,他并没有把自己对于一个事物的理解强加到他的培养客体上,而是去共同发现,用他的话说,研究就是老师和学生共同对一个未知的问题展开互助合作的探索,如果这个问题老师已经很清楚了,那就是讲授,而不是研究。这一段对于教学情境的描述为我们揭开了开放式建筑教育体系的一个侧面。在这里,学生不是被动的接受者,而是探索研究过程中的主导者。

在传统的建筑体系中,老师常常作为一种知识的传授者,而不是引导者,这个时候学生的发展也受到老师知识体系的限制,这就是一个向上封闭的体系,而不是开放的(图1)。

在这里对于开放式建筑教育体系的认识是应该让学生成为这个体系的主体,老师在这其中只是引导者,让学生自己找到他们自己的建筑观,而不是去复制和模仿别人(图2)。

当然,老师在这其中并不是旁观者,他们需要把这一探索过程中所必备的知识和一些获取知识的途径介绍给学生,并让他们在一个相对轻松的气氛中找到他们自己。在这个过程中,老师需要在适当的时候留白,正像我们在作画的时候所做的一样。

2.2 走向多元——建筑教育的思维拓展

走向多元是对于建筑设计教育开放式体系认识的核心转变,这就意味着对于一个设计课程来说应该有一个开放的设计结果和评价体系,这里的重点是培养学生的多方案思维,也就是面对一个设计任务时的开放性思维。一个开放的思维体系和评价体系就包含了对于方案可能性的无限设想,学生应该从一个出发点找到一个设计的若干结果,他们提交的作业也应该是几个方案。在这一体系中,了解一个项目从各角度和各方面利益出发可以有很多种可能性,最终选择一个项目结果的人并不是设计师,而是这些隐含的社会因素作用的结果。开放式的结果将使得设计者更加适应今后的生产体系(图3),这一点将解决现在设计教育中教学和实践脱节的问题。在传统设计教学的过程中往往是由老师担任设计教育的主体和评价者,这就大大限制了学生的思维,形成了一个非开放而封闭的结果,从而让学生的思维越来越窄。

3 双管齐下——培养多元思维的开放型建筑人才

3.1 营造多元环境——开放的教育体制

多元环境的构成有诸多因素:开放的办学思想、开放的学科方向、开放的学生来源、开放的师资队伍等,这是一种硬件概念的开放,通过体制优化可以实现。

在我们的研究项目"开放式研究型建筑设计课程的教学组织"中,我们曾经对这种多元环境的营造进行研究,连续三年在本科生教

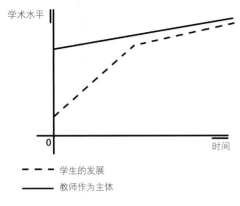

图 1 传统教育模式中的师生关系

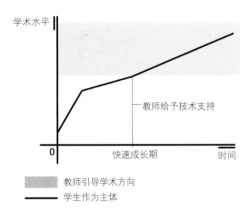

图 2 开放式教育体系中的师生关系

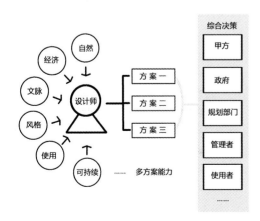

图 3 建筑方案的生成和选择过程

学中聘请法国拉维莱特建筑学院的杜博斯克教授进行为期一周的联合设计，在课程教学的过程中教师对其设计理念和方法进行介绍，并让学生以其设计方法为载体进行方案设计探索，在多方案的比较中进行优选优化，并尊重学生的个性，使其最初的想法得以实现。该课程获得了2011"高科技生态建筑设计"Autodesk杯教案优秀奖、2011年"Gathering above the river"Autodesk杯优秀作业等荣誉。这是对于国外教育力量的引入；除此之外，我们还组织研究生参加国际合作竞赛，培养其国际化视野。其中包含国外设计考察、国外真实设计项目调研、设计讨论、多元设计结果评价等一系列环节，最终也获得了2011年国际建筑合作小组IL PREMIOCOMPASSO VOLANTE竞赛方案一等奖（图4）。

在这一竞赛过程中，每一个研究生都提出自己对于方案的若干种设想，并通过与甲方的沟通和团队讨论取舍大家的想法，最后生成一个具有共识性的方案。在这里，结果仅仅是过程正确性的证明，也为我们优化设计教育体制提供了一个论证。

3.2 培养多元思维——开放的教育理念

培养多元思维需要我们在软件方面革新我们对于建筑设计教育的传统认识，这也是在开放式建筑设计教育理论中最为重要的一个方面，软件的革新是要在设计教育中改变教育者的思维，这一点也需要通过教育本身来实现，这就对开放式建筑教育理念进行了深化。

开放的建筑教育理念要求学生成为主体，老师作为引导者，对一个设计问题进行研究，并得到一个开放式的结果，从而培养学生的多

图 4 组织研究生参加国际竞赛

元思维。这就从设计思维上实现了开放性，而不是流于一个开放的课堂本身。换言之，一个开放的课堂不等于开放教学模式本身，也不代表其一定能用形式的开放培养出具有开放思维的人，这是我们在进行开放式教育研究中必须要意识到的，关于开放式教育研究最关键的一点还是思维的开放，从老师的思维到学生的思维，只有这样才能达到培养具有多元思维的开放型建筑人才的目的。

在一个开放的世界格局中，建筑教育的开放化图景已经展现在我们面前，其新颖的形式和多彩纷呈的设计结果吸引了人们的眼球。可是我们必须认识到开放也有表层和深层两种含义，表层的是体制上的形式化的开放，深层的则是教学思想的开放和设计思维的开放，只有这两个层面相辅相成地发展，才有可能完成开放式建筑教育体制的最终目标：培养具有多元思维的开放型建筑设计人才。

参考文献：

[1] 朱文一，刘健. 面向世界的清华建筑教育[J]. 城市建筑，2011(3).

[2] 栗德祥. 呼唤开放式建筑教育体制[J]. 新建筑，2000(1).

[3] 史建，冯恪. 王澍访谈——恢复想像的中国建筑教育传统[J/OL]. 世界建筑，2012(05):24-29.

建筑设计系列课程开放式评价体制改革探索[1]
The Basic Design Education Structured by Introducing Psychological Quality and the Relative Evaluation Strategies

史立刚　Shi Ligang
周立军　Zhou Lijun
董　宇　Dong Yu
哈尔滨工业大学建筑学院
School of Architecture, HIT

摘　要　本文首先介绍了传统建筑设计评图方式并分析其弊端，然后提出了注重总结性评价与形成性评价相结合、专业评价与大众评价相结合的开放式评价体制改革理念，并结合哈尔滨工业大学建筑学院近年来的开放式评价体制改革实践，提出了联评式、展览式、答辩式、综合式等四种评价模式，最后提出了深化开放式评价体制改革的具体建议。
关键词　哈尔滨工业大学，建筑设计，开放式，评价体制

Abstract　This article introduces traditional assessment of architecture design and analyses its drawbacks, and put forward open-system evaluation concept, which not only focuses on the combination of summative evaluation and formative evaluation, but also the combination of public evaluation and professional evaluation. It combines the reform of open work evaluation in school of architecture, Harbin institute of technology the last years, proposes joint assessment, exhibition, Respondent and Integrated four types of evaluation methods. Finally it brings forward suggestions deepening open evaluation reform.
Keywords　Harbin Institute of Technology, Architecture Design, Open Evaluation, Evaluation System

《国家中长期教育改革和发展规划纲要（2010-2020年）》中强调："深化教育体制改革，关键是更新教育观念，核心是改革人才培养体制。"作为检验教学改革成效的标尺，考试制度改革是至关重要的。改进教育教学评价，根据培养目标和人才理念，建立科学、多样的评价标准。探索促进学生发展的多种评价方式，激励学生乐观向上，自主自立，努力成才。建筑设计系列课程是建筑学本科生的主干专业课。对于学生而言，设计的开放式、建构式学习方式与以前的学习方式反差较大，特别在多样的评价体制面前往往一头雾水，无所适从。对建筑设计课的兴趣如何往往决定了以后职业生涯的态度，因此，如何从各个环节调动学生的热情是我们教学工作的重中之重，而开放式评价体制的改革是教育改革工作的重要组成部分。

1 传统建筑设计评图方式及其弊端

传统建筑设计课作业基本由任课教师根据作业质量、设计过程和课堂表现评定，这种既当教练又当裁判的模式延续了多年。由于建筑设计缺乏统一而定量的评价体系，任课教师的执教经验和对评价原则的理解和眼光偏好又各有千秋，各有侧重，在评价主体一元制的制度下，任课教师集教练与裁判角色于一身，学生的命运仍然掌握在教师一人手里，容易陷入片面的主观评价的混乱局面，这不符合现代开放的教育模式和教育理念，严重地制约了专业教育的推进，不利于青年的身心健康发展，与21世纪课程标准所提到的评价要求相背离，长此以往下去，势必会成为推进高等教育的障碍。同一份作业在各教师间的审阅结果可能差异较大，这样对教师间的交流沟通和对学生的科学公正都是很大的阻碍；同时，各任课教师间的教学成绩会出现各自为政、发展不平衡的局面，而学生也不清楚自己的作品在全年级的位置

1　本文发表于：2011全国建筑教育学术研讨会论文集. 北京：中国建筑工业出版社，2011。

和水平，不利于教学环节的查漏补缺、教学水平的提升以及教学过程的规范。以科学公正的名义改革建筑设计的作业评价机制势在必行。

2 开放式评价体制改革理念

2.1 评价与评分不是同一个概念

过程评价(形成性评价)是在1967年由美国哈佛大学的斯克里芬(M.Scriven)在开发课程研究中提出的。布卢姆（B.S.Bloom）将其引入教学领域，提出了掌握学习的教学策略，取得了显著的成绩。布卢姆侧重于对学习过程的评价，并把评价作为学习过程的一部分。布卢姆主张教学中应更多地使用另一种评价方法——形成性评价或形成性测验。布卢姆指出，评价与评分不是同一个概念。不评分也能够对学习结果作出评价，形成性评价就是一个例子。形成性评价的有效的程序是：把一门课分成若干学习单元，再把每一个单元分解成若干要素，使学习的各种要素形成一个学习任务的层次，确定相应的教育目标系统。形成性测验常常被用来控制学生的学习进程，保证学生在开始下一个学习任务之前，完全掌握这一单元的内容。

形成性评价是对学生的学习过程进行的评价，旨在确认学生的潜力，改进和发展学生的学习。形成性评价的任务是对学生日常学习过程中的表现、所取得的成绩以及所反映出的情感、态度、策略等方面的发展作出评价。其目的是激励学生学习，帮助学生有效调控自己的学习过程，使学生获得成就感，增强自信心，培养合作精神。形成性评价不单纯从评价者的需要出发，而更注重从被评价者的需要出发，重视学习的过程，重视学生在学习中的体验；强调人与人之间的相互作用，强调评价中多种因素的交互作用，重视师生交流。

建筑设计课程评价标准针对过去评价只重视总结性评价而产生的种种弊端，提出总结性评价与形成性评价相结合，更关注形成性评价的新理念。

2.2 多元化开放式的评价理论

科学的评价体系是实现课程目标的重要保障。多元化开放式评价是指评价主体的多元化和评价形式的多样化。它主要是学校、教师、学生之间对学生在知识与技能、过程与方法以及情感、态度、价值观等方面的发展状况的评价。教学评价不单是在学校中进行，而应在整个社会中进行。它应该在相应的情景中进行，应以专题作业、过程作品集等形式架起校内活动和校外活动之间的桥梁。评价方案要考虑到个体之间的差异、发展的不同阶段和专业知识的多样化。评价特别强调对个体智能强项的识别，充分发挥学生个体的潜能。评价的过程应该贯穿在整个学习过程之中，无论是课堂上的学习活动，还是学生参加设计竞赛等实践活动等，均含在评价之内，并注重大众评价与专业评价相结合。

3 建筑设计系列课开放式评价模式探索

通过教学改革与教学研究，本着"公开、公平、公正"的原则，哈尔滨工业大学建筑设计系列课均实行课堂评价、阶段评价与期末评价累加式评价方法，以实现教学和评判的良性循环与反馈。具体而言，包括学生自我评价、教师综合打分和教学组综合联评以及大众评价。在设计的各个阶段，组织学生对彼此的方案进行评价。教师对学生的考查不应局限于最后的设计成果图，而更应注重设计各阶段"过程性"成果的综合评价。由于设计课周期为64课时，历时8周，所以在设计的一草、二草及上板草图中均给出分数，作为平时成绩的依据，并结合集体讲评、课堂辅导、分期检查等方式引导学生改进学习方法，提高学习主动性，强调学生在"过程设计"中的能力提高，避免对设计结果的过分关注，有效提高学生的创新能力和综合素质。课程设计结束后，组织各班级联评。这种多元（评价主体多元化）多价（评价原则、要点多层次化，过程多阶段化，形式多样化）的评价体制加强了评判的透明度，取得了良好的教学效果。

3.1 联评式

联评式是指评价主体从任课教师本人扩展到教研组全体教师以及相关专业教师，而评价客体则从单一的设计图纸扩展到作品装置及多媒体文件。面向全部学生进行开放式的作业联评，形成权威性的评议意见。课程设计成绩分两个部分，平时过程作业占30%，设计成图占70%。在设计分数评定后两周内，对所辅导的班级学生进行总结讲评。这种评价模式的专业性与可操作性均较强，克服了建筑设计作业指标较具弹性、任课教师评价主观片面的弊病。作业集体匿名联评的成绩可多元采样，从统计学角度上看相对客观，可较准确地反映学生作业的真实水平；同时通过共同对作业的评审，增进任课教师间的交流与统一认识，相互之间的监督有利于提高评判的透明度，对于学生相对公平。哈尔滨工业大学建筑学院建筑设计系列课程作业已经全面推行了集体联评，任课教师采用回避制，最终实现教练与裁判的分离（图1）。评完再进行集体讲评和作业展览，使学生对自己的作业与优秀作业之间的差距有更明晰的认识，取得了良好的收官效果。

3.2 展览式

展览式是指评价主体从教研组教师进一步扩展到全体师生，而客

图 1 2009 春季作业集体联评现场

图 3 2008 春季作业二区主楼阳光大厅展览反馈

图 2 2008 春季作业二区主楼阳光大厅展览盛况

体则是受众面更广的模型、表现图等。这种评价模式由于主体更为开放,学生作品直接与潜在使用者接触,使学生对自己作品的社会接受度有更深的理解,对专业更有成就感,其社会影响远远超过了专业评价层面。作为哈尔滨工业大学建筑学院建筑设计课实践教学改革的成果之一,建筑系2008~2010年连续三年对大师建筑名作模型在二区主楼阳光大厅进行公开展览(图2),为师生呈现了栩栩如生的建筑模型盛宴,赢得了过往师生的驻足观看与好评,大众评委的投票占有作品评价指标一定的权重(图3)。这种模式既扩大了建筑学院的形象和知名度,增进了各院之间的交流沟通;从另一个角度讲,对其他学院师生的建筑美育修养也有很大的推动作用,又提高了学生进行课业设计的积极性和主动性,同时也增进了建筑设计课作业评价机制的公开、公正、透明,教学效果获得了广大师生的认可。原国防科工委网站于2008年5月28日也以"哈工大建筑学院引入体验教学,磨炼学生实践'真功'"为题作了报道。

3.3 答辩式

答辩式是指学生对自己的设计理念、构思过程和设计成果进行系统总结,运用多媒体陈述,教研组以及相关教师组成的答辩组针对设计作品进行面对面的交流提问,学生回答问题。这种实战模拟的评价模式使学生与评委直接接触,使设计思想得到更为充分的解读,在提高设计图纸和模型表达能力的同时,也锻炼了口头表达能力,全面考核了学生的综合素养,以期更快地适应社会和市场需求。这种评价模式不仅检验毕业设计，而且也广泛应用于各年级的设计评价过程中。2011年春，哈尔滨工业大学与麻省理工学院联合设计小组经过一个月的分头设计，在4月进行的中期检查中联合应用答辩式评价模式，学生分组进行设计思路和成果的多媒体展示，并进行问答式的英语交流，对于国际联合设计评价是一次有益的尝试。

3.4 综合式

为更好地展示作品的设计理念，学生在设计过程中通过图纸、模型和多媒体与教师进行汇报和交流，通过技能考核、专题活动、自我评价等方式完成形成性测验；设计完成后在现场进行展览，过往师生均能投票、品评，形成大众评价；专业评审时，学生进行讲解、演示和体验，教师评委可以面对面地提问交流，最后结合集体联评图纸的情况给定成绩。2011年春一年级的8m³建构作业的评审就采用了这种方式。综合式评价模式融合了专业评委与大众评委的意见，总结性评价与形成性评价相结合，使建筑设计作品的评价体系更为科学系统，对类似作业的评价也具有一定的借鉴意义（图4、图5）。

4 结语

德国教育家第斯多惠说过：教学的艺术不在于传授本领，而在于激励、唤醒和鼓舞。这句话充分说明了对学生评价的作用，评价也是一门艺术。评价学生所起的是一种激励与促进作用，让不同潜质的学生能通过评价看到自己的优点，培养自己的成就感；同时培养一种团结向上的协作精神，同学之间只有在协作的氛围中才能获得小组评价的好成绩；更重要的是培养学生对建筑设计的兴趣，在一种快乐与渴望中学习，渴望与同学之间的竞争与合作。因为开放性的评价可以是多方位、多角度、多层次、多元化的，教师和学生可将建筑设计的学习评价延伸至课堂内外，以此来不断地激励学生进步。评价体系因子和指标的模糊性是开放式评价体制的一大难题，绝对标准化的评价体系不可行，制定详细而操作性较强的评价细则并高效执行是我们的出路。

教育改革的一个关键性环节是考试制度改革，目前来看任重而道远，建筑学人仍需努力探索。分数不是关键，但它具有一定的指向作用。90后的青年学生个性更加突出，主体意识较强，哈尔滨工业大学建筑设计系列课程适应这一时代特点，通过近年来的开放式评价体制改革实践，初步摸索出了较具可操作性、师生互动和认可度颇高的评价体系，在今后的教学改革深化过程中还需要进一步完善，希望我们的经验能为兄弟院校建筑设计及相关课程的评价提供借鉴。

参考文献：

[1] 形成性评价——学生发展不可缺少的评价方式.

[2]（德）第斯多惠. 德国教师培养指南. 袁一安译. 北京：人民教育出版社，2001.72-74.

图4 2011春8m³建构作业综合评价现场（一）

图5 2011春8m³建构作业综合评价现场（二）

Integrating Science and Technology into Design Teaching[1]
将科学与技术融入设计教学

Jian Kang
School of Architecture, University of Sheffield, Sheffield S10 2TN, UK
School of Architecture, Harbin Institute of Technology, China
康健
谢菲尔德大学建筑学院，谢菲尔德 S10 2TN，英国
哈尔滨工业大学建筑学院，中国

Abstract The teaching philosophy of integrating science and technology into architectural design is explored and demonstrated through the discussion of the teaching structure, student submissions, and examples of scientific visualisation tools at the School of Architecture, University of Sheffield.
Keywords Architecture, Science and Technology, Teaching, Design, Integration, Scientific Visualisation

摘　要　将科学和技术融入建筑设计的教学理念是通过英国谢菲尔德大学建筑学院的教学结构，学生意见以及科学可视化工具的例子来探索和论证的。
关键词　建筑，科学和技术，教学，设计，集成，科学可视化

1 Introduction

The School of Architecture, University of Sheffield is one of the largest architecture schools in the UK. Within a research led university, the school has a strong emphasis on research-led design and it also has the best Research Assessment Exercise (RAE) track record among UK architecture schools. The School was combined from the original Architectural School and the Building Science Department. As a result, teaching and research in sustainability and technology is a notable strength of the school.

The teaching philosophy for science and technology at the school is to integrate it into architectural design, which is explored and demonstrated in this paper.

2 Teaching content and structure

At the School of Architecture, University of Sheffield, Science and Technology is taught in various courses, including:

(1) The Royal Institute of British Architects (RIBA) accredited architectural courses, namely

o RIBA Part I (3-year undergraduate course leading to bachelor degree, i.e. Year 1-3, followed by a year-out in practice as Year 4);

o Part II (2-year course leading to diploma and/or master degree, i.e. Year 5-6);

(2) Various joint courses for dual degrees between architecture and landscape, between architecture and urban planning, and between architecture and civil engineering;

(3) One-year taught master courses, in sustainable design and in other subjects; and

(4) Research degree programmes including Master of Philosophy (MPhil) (1-2 years) and Doctor of Philosophy (PhD.) (3-4 years).

1　本文为"2011中英韩建筑教育国际论坛"宣读论文。

For the architectural courses and the dual degree joint courses, typically 120 students each year in total, the teaching philosophy for Science and Technology at the school is to integrate it into architectural design.

Science and Technology is taught throughout the duration of the courses. For example, acoustic modules are taught in:
(1) First year (basic room acoustics and noise control);
(2) Third year (advanced room acoustics and noise control);
(3) Fifth year (acoustics and noise control towards practice including building regulations in acoustics), and
(4) Sixth year (special study and technical report).

For the undergraduate courses, in each year there are two parts:
(1) In the autumn semester lecture courses are given and this is assessed by written exams, and
(2) In the spring semester the technical teachers will go to

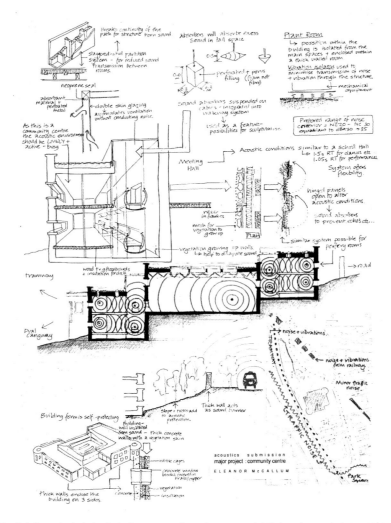

Fig. 1 1st year technical submission on acoustics, where some details for a house design are shown.

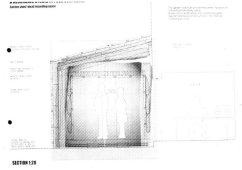
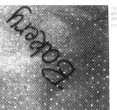

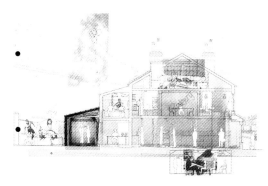
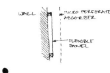
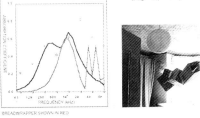

Fig. 2–1 An example of the technical submission from a student in the 6th Year, where the aim is to use low cost, recycled materials to design recording studio.

Fig. 2–2 An example of the technical submission from a student in the 6th Year, where the aim is to use low cost, recycled materials to design recording studio.

Bachelor/Master	Course Title	Type	Place in curriculum	Teaching methods	Content
Bachelor in Architecture & Dual degrees with architecture	Science & Technology I	Compulsory	1st Semester Year 1	lectures	Basic room acoustics and noise control
			2 Semester Year 1	studio hours	
	Science & Technology III	Compulsory	5th Semester Year 3	lectures	Advanced room acoustics and noise control
			6th Semester Year 3	studio hours	
Diploma in architecture (master level)	Science & Technology V	Compulsory	9th Semester Year 5	lectures/workshops	Acoustics and noise control towards practice
			10th Semester Year 5	studio hours	
	Science & Technology VI	Compulsory	11th Semester Year 6	lectures/workshops	Special study and technical report
			12th Semester Year 6	studio hours	
Taught master MPhil/PhD	Environment simulation	Compulsory	1st Semester	½ lecture + practice	Acoustic simulation

Table 1 Teaching structure of Science and Technology at the School of Architecture, University of Sheffield.

Institutions	Bachelor		Master in architecture (unspecialised diploma)	
	Place in curriculum	Compulsory application to design project	Place in curriculum	Compulsory application to design projects
National High School of Architecture and Landscape of Bordeaux in France	2nd, 3rd and 5th semesters	+	3rd semester (year 2)	+
Second University of Naples in Italy	5th semester	-	5th and 6th Semesters	only to final project
Sheffield University in the UK	1st, 2nd, 5th and 6th semesters	+	9th, 10th, 11th, 12th semesters	+
Yıldız Technical University (YTU) in Turkey	5th semester	-	-	-

Table 2 Comparison between four European architectural schools, in terms of acoustics/noise control teaching.

studios to help students with their design schemes, and this is assessed by technical submissions, which also form a part of the design portfolios.

In Table 1 the teaching structure of Science and Technology is summarised. It is noted that for the third year and the sixth year, some students also choose Science and Technology as the topic of their thesis. In Table 2 a comparison is made between four European architectural schools, in terms of the acoustics and noise control teaching, as an example. It can be seen that the model used in Sheffield encourages the integration of such teaching into design.

In the UK, students enter universities by passing three subjects at high school level (A-level). The backgrounds of architectural students vary considerably, with different strengths in science, social science and arts. It is therefore rather challenging to teach in a way that every student can benefit. Integrating Science and Technology into design, the common theme of all students and the focus of the courses, is thus of great significance.

The MPhil. and PhD. courses are by research only in the UK. For example, the Acoustics Group currently consists of over 10 PhD. students with a range of backgrounds including acoustics, architecture, planning, environmental science/engineering, computer science and sociology/psychology. The Group has a range of acoustic facilities and measurement equipment, as well as simulation software. The topics of the research students cover a wide range, including room acoustics, building acoustics and environmental acoustics.

3 Scientific visualisation tools

It is important to develop simplified scientific visualisation tools to aid design. A series of such tools have been developed and used at the School of Architecture, University of Sheffield. The principles of the tools are to consider the following factors:
(1) Effectiveness of various key parameters;
(2) Scientific visualisation way of presenting teaching materials;
(3) Interactive and user-friendly;
(4) No extra learning time will be required for using the tools, with basic knowledge taught in the lecture course.

Areas covered include urban noise, room acoustics, and sound materials. Both classic formulae/models and some state-of-the-art techniques are considered. The long-term objective is to develop a complete set of tools.

The tools include:

(1) Sound distribution behind an environmental noise barrier, where the parameters include barrier height, source-barrier distance, source height, and receiver positions. This is shown in Fig. 3.

(2) Sound distribution in a rectangular street canyon, where the parameters include street length, width, building height, boundary absorption coefficient, air absorption, and receiver positions. This is shown in Fig. 4.

(3) Reverberation time calculation in a rectangular space, where the parameters include room dimensions and boundary absorption coefficients. This is shown in Fig. 5.

(4) Absorption of micro-perforated panel absorbers, where the parameters include hole size, hole spacing, panel thickness, and depth of airspace. This is shown in Fig. 6.

(5) Digital audio animation for urban soundscape design, where the tool would allow users to input idealised cross-streets and squares in a 2D environment, and after selecting and positioning a number of urban sound sources, the soundscape file with multiple sources can be played back, with reverberation effects. This is shown in Fig. 7.

4 Conclusions

Through the above discussion and analysis, it is clear that the following points would be of great values:
- Science and technology as a key part of teaching
- Teaching throughout different levels
- Teaching throughout courses
- Integrating technology teaching into design
- Using scientific visualisation tools
- Considering students background
- Link teaching and research, such as through special studies and technical reports, or through links with research projects, or through using research students in teaching.

Acknowledgements

The development of the scientific visualisation tools was

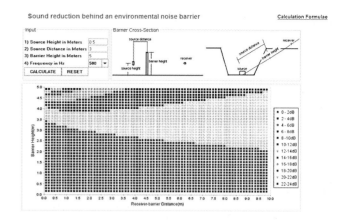

Fig. 3

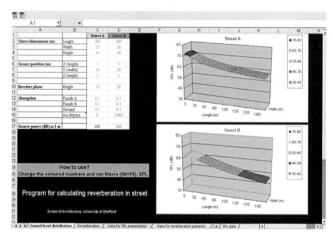

Fig. 4

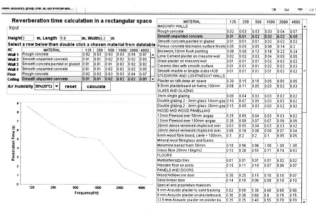

Fig. 5

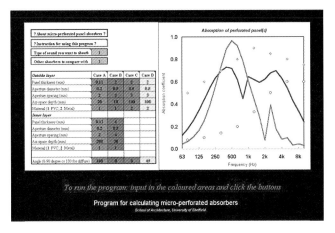

Fig. 6

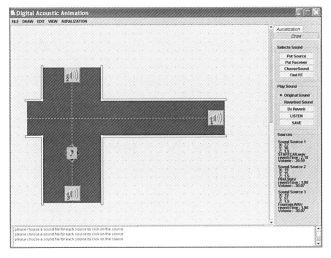

Fig. 7

financially supported by the USA Theodore John Schultz Grant - For Advancement of Acoustical Education, and also by the Hong Kong TDG. This work is also partly funded by the China National Science Foundation (50928801).

References:

[1] Can Z., Maffei L., Semidor C., and Kang J. Experiences in noise control education in architecture. Proceedings of the 38th International Congress on Noise Control Engineering (Inter-noise). Ottawa, Canada, 2009.

[2] Kang J., Tsou J. Y., Peters T. F., Hall T., and Lam S. Integrating scientific visualisation into architectural curriculum for teaching acoustics and noise control. Architectural Education Exchange - Teachers in Architecture: Facing the Future. Sheffield, England, 2000.

[3] Kang J. Computer tools for architectural acoustics education. Proceedings of the Acoustics'08 Paris, France, Also published in J. Acoust. Soc. Am., 123(5), Pt. 2, 3653-3654 (2008).

培养卓越建筑师的实践与创新研究 [1]
The Research of Training Excellent Architects' Practice and Innovation

孙伟斌　Sun Weibin
路郑冉　Lu Zhengran
哈尔滨理工大学建筑工程学院
School of Architecture and Engineer,
Harbin University of Science and Technology

摘　要　本文通过总结哈尔滨理工大学建筑学专业的教育实践，研究卓越建筑师的培养模式，建立规范化的人才培养方案，确定有效的教学方法和评价体系，提高未来建筑师的创新能力和职业素养。
关键词　卓越建筑师，实践，创新

Abstract　According to summary the education practice of School of Architecture & Engineering, Harbin University of Science and Technology, this article is about the research of training mode of excellent architects, the establishment of the standardization training scheme of the excellent architect, the determination of the effective teaching method and evaluation system and improvement of the future architect's innovation ability and professional quality.
Keywords　Excellent Architect, Practice, Innovation

1 引言

中国三十多年的改革开放，带来了经济的迅猛发展和巨大的建设规模，每年都有大量的设计任务需要完成，而成熟的建筑设计人才又相对短缺，因此要求建筑院校培养的毕业生能很快地担当重任。对我国目前的建筑教育调研发现，一些院校的建筑学本科毕业生在不同程度上存在一些共性问题，如设计概念上缺乏独立的评判精神和深入思考分析的能力，建筑的总体设计、功能分区、交通流线和空间设计等方面的掌控能力不够，甚至有些学生对办公楼、商业建筑和住宅等常见类型建筑和结构的基本原理也不甚明了。为适应工作的需要，设计机构实际上承担起了对毕业生进行职业训练的任务。但这种方式存在着一些弊端：一是各设计机构的职业训练不规范，良莠不齐，对建筑师的成长不利；二是毕业生参加工作后本应通过简短的适应期就可以开始工作，但这种二次教育延长了建筑师的成熟期，造成了设计机构的资源浪费，这与我们所提倡的建设节约型社会是完全背道而驰的。

中国"卓越工程师教育培养计划"拟用10年时间，培养百余万高质量的各类型工程技术人才，为建设创新型国家、实现工业化和现代化奠定人力资源优势。本研究致力于探讨卓越建筑师的培养模式，确定有效、可行的教学策略、途径和措施，建立规范化的建筑学人才培养方案，应用于教学实践，将有助于建筑学专业教学的科学化，促进未来建筑设计人员具有更好的创新能力和职业素养。这对提高我国建筑学教育的质量和加快建筑学学科建设的步伐具有深刻的现实意义。

2 卓越建筑师培养实践

2.1 教学课程实践

在"卓越工程师培养计划"之前，哈尔滨理工大学的建筑学专业一直以"职业建筑师"为培养目标，在课程设置上强调学生实践能力的培养。学生每学期都有实践课程，使职业教育贯穿始终，主要内容有：

1　本文为"2011中英韩建筑教育国际论坛"宣读论文。

第二和第三学期安排学生的钢笔画写生和水彩渲染，提高学生的专业基础技能，为以后的设计打下良好的手绘功底；第四学期进行建筑工地和设计院的认识学习，建筑工地实习使学生了解学习专业知识的目的以及各部分专业课程的相互关系，避免盲目学习，设计院认识学习，带领学生参观知名度较高的设计院，如哈尔滨工业大学建筑设计研究院、黑龙江省建筑设计研究院等，体验式教育和必修、选修课程，让学生在学习专业知识之余，了解社会职业环境，探索个人职业方向；第五学期和第六学期进行结合计算机的表现技法训练和测绘实习，提高学生的计算机表达能力，为后续的设计实践环节打下良好的基础；第七学期进行调研实习，通过参观各地的古建、优秀建筑以及城市规划成果，使学生开阔眼界，避免闭门造车；第八学期安排了针对设计院实习的业务实践培训；第九学期进行建筑师业务培训，学生选择学校的实习基地或国内各地的设计院实习，在真实的环境中体验真实的工作流程；第十学期毕业设计，通过综合性的建筑设计，使学生能综合运用所学知识，定向指导使学生在现实与理想中明确职业方向，为未来的工作做好准备。另外，从第一到第七学期，还穿插着聘请全国著名的建筑师进行形势与政策讲座，安排多个不同规模和时间要求的快速设计。

通过对建筑学专业2005级和2006级两届毕业生的跟踪调查以及用人单位对实习生和毕业生的信息反馈，大部分学生专业素质较好，工作上手较快，得到了用人单位的认可。今后我们将继续完善建筑学本科专业的实践性教学体系，加强和细化实践教学环节，使职业建筑师教育贯穿于建筑学教育的全过程，通过全程化专业指导和分层递进式的职业导航服务，让学生更好地规划职业生涯，以培养更多的卓越建筑师。

2.2 教学培育计划的针对性调整

2010年，建筑学专业定点为哈尔滨理工大学"卓越工程师培养计划"的实验专业。根据建筑学专业特点以及培养卓越建筑师的需求，我们分析总结了2005、2006版培养方案运行的经验教训，在此基础之上制定了2010版培养方案，提出了"厚基础，重实战，培养卓越职业建筑师"的培养模式。

2010版培养方案的特色是：突出卓越建筑师的培养体系，增加有关国家注册建筑师培养大纲要求的内容，在培养方案内做到全覆盖，并重点突出"创新能力、工程设计能力的培养"；突出建筑学专业学生创新性培养，将技能训练、课程设计等实践、实验环节与设计竞赛和大学生开放性、创新性实验相衔接，进一步培养和提高学生的创新能力、动手实践能力，目前已经取得初步成果。

2010版培养方案的主要调整内容是：根据市场、企业用人单位的反馈，调整了部分课程的学分、课时和教学时序，使教学体系更加紧凑和顺畅；加强卓越建筑师规划教育，将原有《建筑师职业教育及法规》整合了职业生涯发展的相关内容，升级成为《建筑师职业规划》；对个别实践环节进行整合，使之运行更顺畅。新一轮的建筑学专业培养方案，使建筑学人才培养以"卓越建筑师培养"为核心，通过合理安排理论与实验、实践环节相结合，提高学生的实践创新能力，使整个课程体系更具有系统性和逻辑性。

3 卓越建筑师培养创新

3.1 教学方法更新

（1）开放式教学

建筑学是一门应用学科，不能仅仅是学院式教育，必须走向社会，开门办学。为此我们采取了一系列方式，如聘请一些大型设计院的有经验的执业建筑师、工程师来讲课，因为他们了解建筑师需要什么样的知识，能教给学生们最实用，最直接的知识，使学生能够"站在巨人的肩膀上远望"。我们还聘请一些参加工作2~3年的青年建筑师参与指导学生的课程设计，这些青年建筑师刚刚离开校门参加工作，既充满激情又感触颇深，在教学中能很好地指导学生课程设计，同时也缓解教师短缺的情况，减轻主讲教师的工作量，使主讲教师可以拿出更多精力，投入到教学研究及科研当中。另外，我们还聘请国内外的知名学者来举办讲座，使学生能了解到理论前沿与研究动态。

（2）培养社会实践能力

学生的社会实践能力对其日后的工作学习尤为重要，因此应引导和鼓励学生参加多种实践如参观、实习以及调研等，而教师也应为学生提供各种各样的实践机会。

课程设计题目强调以"真题"为主，真实的环境和市场需求，贴近实战演练，激发了学生的设计热情，在"做"中学。使学生在真实可调研的地段，学习执行相应的法律法规，在课程设计中创造性地学习。如2006级学生的城市设计项目采用的是实际项目，题目为"哈尔滨市道外历史文化风貌保护与再利用设计"。项目地点位于哈尔滨市道外区中心区，由南五道街、靖宇街、南头道街以及南勋街围合而成，总用地10.3公顷，该地块是哈尔滨道外历史保护开发的二期工程。课程设计要求学生从哈尔滨老道外历史文化街区的现实问题出发，通过现状调研与分析，寻求科学合理地进行历史文化街区保护与更新的方法，采用整体渐进式的有机更新模式，从保护控制分区的重构、目标控制体系的建立、多元文化内涵的拓展、街区用地的整合利用等方面实现整体街区复兴的对策，并协调保护与开发之间的关系，促进哈尔滨道外历史文化街区的可持续发展。在真实设计中，3~4人一组，每组的调研汇集到一起，在合作中自主学习和安排学习进度，准确定位以发挥优势和提高合作效率，在协作中学习别人的长处、弥补自己的不足，培养良好的团队精神。

另外，通过成立导师工作室，提供实际项目鼓励学生参与，联系多家设计院给学生提供假期实践实习地点等，通过一系列学生主动参与的调研与实践活动，丰富了学生知识的同时也教会了他们如何去学习、如何去做设计，无论今后遇到什么样的设计难题，他们都会从容应对，因为这种继续学习、勇于创新的能力将伴随他们的一生，而这才是大学教育教授给学生的最宝贵的财富。

3.2 评价体系更新

（1）评价主体的更新

目前，我国各高校建筑学作业评价主体是任课教师，这些教师大多是"从学校进，到学校出"，虽然有少数具有实践经验的执业注册建筑师调入高校，但也因为高校制度的限制而使他们再鲜有机会接触到后方案阶段的设计，也渐渐趋同于其他教师。这种单一的作业评价主体不能及时地获得最新信息和技术反馈，与生产实践脱节，助长了建筑学专业学生只重表现，不重技术的不良风气。据来我系讲学的英国谢菲尔德大学建筑系康健教授介绍，他们每年都聘请建筑事务所中成熟的建筑师到各个年级授课，效果非常好。今后，我们要打破常规，将评价主体由单一的任课教师，扩展到任课教师、从业建筑师、学生群体及其他参与该设计的相关人员（如区域内居民、建设方、承建方、规划主管单位等）。这种多元评价主体无论对学生还是任课教师都具有与以往不同的意义。

在一些实践课程中，评价主体除了教师之外还加入了开发商、政府主管部门和相关合作设计院，使得学生的作品得到了市场的检验和评价。

（2）评价方式的更新

传统的封闭式评图缺乏透明度和公正性，学生往往不清楚成绩的取得依据，难以在以后的学习中得到改正和提高，另外，容易对成绩产生疑义，挫伤其学习的积极性。设计实践课程采取了以下改革措施：

①展评结合。以往的成果评价环节都是：成果上交——教师评定——成绩下发，学生、成果、教师之间处于相对隔离的状态，难以产生沟通和交流。现改变为设计成果全部公开展示，既加强了学生之间的相互学习，也加强了对教师所给予的成绩评定的监督。

②讲评结合。改变传统封闭的评图方式，全部班级集中进行课堂公开挂图评，学生以答辩的形式简介其方案构思及特色，同学当场提问，教师现场点评，给出成绩，将封闭的评图过程变为生动活泼的师生互动式课堂教学，既保证了成绩评定中的公正性，又锻炼了学生的表达及应对能力。

③教评分离。改变以往完全由任课教师评图的方式，从各组抽调教师组成评图委员会（教师逐次轮换），主讲教授轮流主持，各评图教师现场独立给分，体现了公平与公开的原则，避免了盲从与偏袒的现象。最后，逐步实行教课者与评图者完全分离。

（3）评价标准的更新

虽然"大作业"式的设计成果的成绩评定存在着主观性和不易量化，通过评价体系的改革，采用化整为零的办法，细化、量化考核各个教学环节，同时给不同评价主体以不同的权重来进行评价，最终可以核定出相对比较客观、公正的成绩。

4 结语

作为建筑学人才培养基地的高校，兼顾着历史和社会的责任而任重道远。建筑学专业学生除了应具有宽厚的建筑理论基础、系统的建筑专业和相关知识之外，还应掌握建筑科学的研究方法和实验技能，并把建筑师的社会责任贯彻到教学中，培养出真正为社会所需的卓越人才。因此，卓越建筑师的教育应用于教学实践，能促进建筑学专业教学的科学化，促进未来建筑设计人员具有更好的创新能力和职业素养。这对提高我国建筑学教育的质量和加快建筑学学科建设的步伐具有强烈的现实意义。

参考文献：

[1] 孙伟斌. 建筑教育呼唤职业建筑师的培养. 2007国际建筑教育大会论文集, 2007.

[2] 孙伟斌. 城市设计课程中应用型人才教育实践. 2010建筑教育大会论文集, 2010.

[3] 蒋方. 关于建筑学专业本科设计作业评价的再思考[C]. 建筑教育论文集. 2008.

Rethinking Professional Skills:
Architects of a Deeper Green[1]
职业技能的重新思考:"更加绿色的建筑师"

Dr. Cristina Cerulli
School of Architecture, University of Sheffield, UK
Dr. Cristina Cerulli
谢菲尔德大学建筑学院,英国

Abstract Despite the prominence nowadays given to sustainability within any built environment curriculum, outside the pedagogical discourse arena,the praxis of teaching and learning sustainability within schools of Architecture is too often limited to environmental design. Whilst most schools of Architecture in UK have, at least to some extent, embraced environmentally conscious design in their curricula,little has been done to promote a shift in the profession towards a more integrated view of sustainability that looks holistically at the environment, including society, and challenges established professional norms and praxes. This paper will discuss 'green management' work in the area of social enterprise and mutual models of development pioneered in the post-graduate architecture curriculum at the University of Sheffield between 2005 and 2011.
Keywords Professional Skills, Pedagogy of Sustainability, Architectural Pedagogy

摘 要 尽管可持续性在当今任何建筑环境类课程中都处于重要的位置,而在建筑教育的论文范围之外,建筑学院中可持续性的教学和学习却过多地局限于环境上的设计。当大多数英国的建筑学院至少在某种程度上把有环境意识的设计汇集在他们的课程中的时候,几乎没有学校在专业上能向着更富有整合性,并包含社会和具有挑战性的专业基准和实践的可持续观点上继续往前迈进。本篇论文将讨论在2005~2011年期间社会性的企业和与谢菲尔德大学研究生建筑课程共同首倡的所谓的"绿色管理"工作。
关键词 职业能力,教育方法可持续性,建筑教育

"In the US there is less visible social activism, and especially in architecture schools and architectural discourse, the notion that architecture has any role to play in the political arena is dismissed. How often I have heard: 'We're architects. We make form. It's up to sociologists, economists and politicians to figure out the rest'. I believe that this is absolutely wrong." (Roberta Feldman in Roberta Feldman & Henry Sanoff 2008)

In the thirteen page long document of Validation Criteria published by the Royal Institute of British Architects (RIBA) (Anon 2002), the word sustainability is mentioned once, as a 'reason', on par with 'budget', to give appropriate design responses to site and context . In the recently revised RIBA Validation criteria, sustainable and sustainability are mentioned three times and always in generic terms (Anon 2011, p.52, 53, 56). Despite this limited institutional push from the accrediting bodies to embed sustainability into Higher Education architecture curricula, however, the real and tangible environmental urgencies around climate change and peak oil have pushed the 'sustainability agenda' into a prominent position in those curricula. The filtering down of sustainable environmental requirements into policies and statutory requirements have also pushed sustainability into the lexicon of mainstream practice, where sustainability is a badge that no

---

[1] 本文为"2011中英韩建筑教育国际论坛"宣读论文。

practitioner can afford to be without.

But what does teaching and learning sustainability really mean?
There is a growing body of literature on the pedagogy of sustainability within and across disciplines. Dyer argues for an approach to teaching sustainability which assumes that learners construct their own concepts of the environment, distinguishing between discipline-based teaching in universities and the more holistic Green Education (Ken Dyer 1997). Warburton adds that, to learn about sustainability, 'deep learning' needs to occur and that the challenge of teaching sustainability is 'to create an active, transformative process of learning that allows values to be lived out and debated, and permits a unification of theory and practice' rather than simple transmission of concrete facts about the environment(Warburton 2003).

Yet I would argue that, when dealing with sustainability, a large number of architecture courses are still operating predominantly from a functionalist perspective, where the 'problem of sustainability is largely technical' (Porter & Cordoba 2009). Whilst Porter and Cordoba maintain that a functionalist perspective is sometimes appropriate to deal with some aspects of sustainability, students also need to develop skills and learning strategies that are Interpretive, involving awareness, appreciation and ethical action and from the perspective of Complex Adaptive Systems, where learners are aware of the interconnectedness of the interdependence of networks of both the subject they are studying and the system they are part of (Porter & Cordoba 2009).

This paper will examine curriculum developments implemented within the Management and Practice modules in the March course at the University of Sheffield, to foster a holistic approach to sustainability that is rooted in social innovation and empowers, enables and exhorts students to become proactive within their environments.

The context of this curriculum development is the School of Architecture's pioneering pedagogical work (Worthington 2000; Torrington 2000; Rachel Sara 2000; Nicol & Pilling 2000; Morrow 2000; Fisher 2000; Chiles 2000; Parnell 2004; Doidge et al. 2000) and the work of the AGENCY - 'Transformative Research into Architectural Practice and Education'-a research centre at the School of Architecture in Sheffield. The AGENCY research group emerged from the alliance of staff and researchers working in and around the subject of architectural practice and education, taking a critical view of normative values and standard processes with an ambition to propose alternatives.

"AGENCY's is concerned with education and research which address new models of architectural practice to confront the big social and political questions of the future. Such models need to be more collaborative, participative and ethically driven and address the social and political responsibility of the architect in a period of rapid global environmental and economic change" (AGENCY 2010).

The post-graduate professional practice architecture curriculum was redeveloped introducing concepts of social enterprise, mutual models and creative financial planning to foster the emergence of a new generation of architects and designers able to tackle social and economic sustainability as well as an environmental one (Cristina Cerulli 2012a).

Professional management courses are still a rather unusual place, within Schools of Architecture, for engaging with sustainability. One of the aims and drivers for the curriculum redevelopment of the Practice and Management modules was to provide opportunities for developing entrepreneurial skills in a context that was aligned with students' values. Despite recent renewed interest towards entrepreneurship in design within higher education (Anon 2007) and the affirmation of entrepreneurial education and activity as a way for young designers to acquire competencies which strengthen enabling them to innovate and to forge their own modes of practice within the wider spheres of design production (Jensen 2004), enterprise is still not an area that architecture students immediately feel an affinity with (or an interest for). The decision to focus the curriculum on social enterprise, mutual models and creative financial planning was initially an attempt to introduce enterprise skills in the curriculum, whilst remaining in line with pedagogical ethos, School values and students' interests, in the context of a free education(P. Freire & A. M. A. Freire 2004). It quickly became apparent that this alignment yielded a powerful way of conceptualizing intrinsically and deeply sustainable design proposals. Whilst remaining within the framework of professional accreditation requirements, the Management and Practice curriculum was radically transformed with an emphasis on the development of those entrepreneurial skills that would empower students and graduates to conceive and implement their own praxes in a way that is aligned with their ethos and value systems.

Central to this curriculum development, praised be examiners, validating boards and students alike, is the idea of working in a meaningful and ethical way with others across distributed networks of actors, rather than in isolation. The relationship of individual actors with the networks they operate within, particularly networks within built environment, or economic networks, is recognized as key to the understanding of contemporary entrepreneurship practices. "What might a prototype network entrepreneur be like, if not the lone economic hero, manipulating networks to his/her own ends?" (C. Cerulli & Holder 2009)

Existing concepts of the entrepreneurial character are tied up with outdated stereotypes that are still redolent of the 'self-made business man' of the Industrial Revolution, or the 'garage geek' of the dot com bubble; there are, however, indications of a shift towards a more positively and ethically tinged type of entrepreneur; one whose socially and environmentally conscious work thrives on true collaborations and creatively and resourcefully assembled systems. Hyrsky and Tuuanen's carried out a study of the language and the conceptualising of entrepreneurs looking at international concepts of entrepreneurial character; they began by describing four 'stereotypes' of entrepreneurial character (the 'Classical', the 'Craftsman', the 'Opportunistic' and the 'Research and Development' entrepreneur) and then conducted a survey focusing on the attitudes to entrepreneurship and entrepreneurial characteristics, where respondents were asked to give metaphorical descriptions of entrepreneurs and entrepreneurship (Hyrsky, K. & Tuunanen, M. 1996). In this survey they found a shift away from the 'stereotypes' originally characterised; respondents were able to give responses "freely and of their own accord" and their responses ranged widely from more predictable images of "Adventurers and Warriors", to more unusual natural images - "an amoeba", an "oak tree" - to a "melting pot" or "lubricating grease" (Hyrsky, K. & Tuunanen, M. 1996). Hyrsky and Tuuanen's study concludes that the term 'entrepreneur' is loaded, and different users have imbued it with a range of different meanings, which have changed over time from previously a pejorative view of "greed, selfishness, unscrupulousness, [and] disloyalty to employees", to a more positive view of the entrepreneur as someone who is "innovative, giving, constructive, [has/takes] initiative" and is a "builder" (Hyrsky, K. & Tuunanen, M. 1996). It is these 'complementary' and feminine entrepreneurial characteristics that are beginning to emerge in increasingly networked modes of practice–taking initiative, but also giving support or advice to others, building strong links among individuals or organisations in order to promote innovation and positive change, co-operative approach and intuitive and creative approach to solving problems. Lovink and Schneider (as cited in (Burke & Tierney 2007, p.58) present the idea of the network paradigm in design as useful way to bypass the preoccupation with end product, and investigate the 'everyday' nature of workings between actors, materials and places:

"The networking paradigm escapes the centrality of the icon to visual culture and its critics and instead focuses on more abstract, invisible, subtle processes and feedback loops. There is nothing spectacular about networking."

The various components of the management courses focus, over two years, on working with others (listening, teamworking, consultation and participation techniques, interaction and engagement with clients and users), briefing and design of a social enterprise (Year 5) and, in year 6, procurement, statutory frameworks, inclusive design, health and safety and whole lifecycle costing. Key to the success of this teaching and learning programme has been the integration of the management teaching with the design work. This has resulted in higher student engagement with management related issues and, gradually, in the embedding of certain concepts explored within the course (mutuality, cooperation, social enterprise) into sophisticated design proposals that, I would argue, are intrinsically more sustainable than those produced without such 'green' management approach.

I will use extracts from work of one student, Jordan J. Lloyd, Jay, to illustrate of the type of integrated thinking that emerges within the course(Jordan J. Lloyd 2009). A systematic description of the course structure, content and activities is addressed in a forthcoming booklet(Cristina Cerulli 2012b).

The Waste Industry & The Rother Valley

Advanced Waste Economy is a design framework for the assimilation of a new manufacturing sector in the Rother Valley based on waste conversion. Over the next one hundred years, the framework aims to consolidate several local nature reserves and country parks into a single entity through a managed ecological restoration strategy. As this new ecology evolves, an active industry develops within the Rother Valley that converts South Yorkshire's waste into 'upcycled' commodities using Cradle-to-Cradle principles[1]. Using peer-to-peer networking practices as a starting point, the project aims to reconfigure our current perceptions of how organisations of all scales can act as nodes of (fairly) equal value with a greater synergetic effect. Furthermore, A.W.E is committed to changing perceptions about waste as a disposable problem that can be ignored, to a valuable commodity that may become a new livelihood for people living in the Rother Valley. Key to the architectures of the Advanced Waste Economy is to combine programmes of waste with leisure activities as a way of changing perceptions towards waste. As design projects sit within the overall framework, in one example, a proposal for an integrated plastic conversion facility, which allows plastic bottles to be converted into biodegradable plastic pellets[2]. These pellets can then be re-manufactured into a range of commodities ranging from biodegradable golf balls to bottled water or perishable food packaging. Such a waste processing facility could paradoxically contribute to increasing biodiversity: for example, if the manufacturer were to place seeds in the core of the ball, the result would be that people hitting balls on an onsite driving range contribute towards biodiversity as the golf ball degrades (providing nutrients for the soil) and the seeds are germinated or carried away by birds.

Fig. 1 Project Summary Statement (Jordan J. Lloyd 2009)

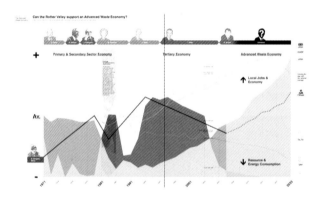

"This drawing charts a Sheffield miners quality of life, indicated by the typical wage within the broader context of political and economic spheres. Sheffield in particular has a labour deficit signalled by the end of the labour strikes and the move towards a serviced based free market system brought in by the Thatcherite Government. Social problems brought on by the collapse of the mining communities (in addition to the mechanisation of other industrial processes such as steel) are ongoing, with current reskilling programmes ineffectual in many cases. An intention to move towards an Advanced Waste Economy is proposed, through technological innovation that currently exists and made commercially viable with increased public awareness of our current waste production. A highly skilled labour force is required to mine, sort, process and disassemble our current waste located in landfill would utilise many of the same skills used in the traditional mining industries" (Jordan J. Lloyd 2009)

Fig. 2 Charting the life of an industrial worker within political change & projections for the A.W.E.

Jordan's project, Advanced Waste Economy (AWE) is a final year thesis project and the passages below are extracts for the Management Report, submitted as assessment for the Management and Practice module. The brief for the assessment is to write "a critical essay, reflecting on Management and Practice issues in relation to and in the framework of your major design project. Your scheme should be used as a vehicle to reflect on generic Management and Practice issues". In the report Jordan uses a combination of short text and diagrams to convey the complexity of his scheme in a way that is accessible. The report opens with a concise and comprehensive statement about the scope, objectives, tools and the ethos of the project (Fig. 1). A thorough analysis of the economical and political context, local, regional and national, is then presented with reference to how AWE would address specific issues. For instance a diagram charts the quality of life of an industrial worker in relation to political change and in the projections for the A.W.E (Fig. 2); another series of diagrams, like Fig. 3, maps the compatibility of core skills of the workforce to in the area showing in detail how this could be redeployed and another diagram maps the 'tag cloud' of stakeholders, where level of influence is represented by size of circle (Fig. 4). Various diagrams explore potential funding and procurement strategies; Fig. 5, in particular, shows how an established procurement method distributes risks and profits between a number of enterprises.

Exploring AWE from the point of view of economics, procurement, engagement with stakeholders was an important moment in Jordan's design process and not a separate exercise imposed by the curriculum. Whilst the quality of Jays work was above the average of the quality of work produced by his cohort, 'green' management concepts are beginning to become consistently embedded in design proposals within MArch design studios.

The impact of these integrated learning experiences is also visible beyond the academic design proposals within which

they occur, in the professional praxes that some of the graduates are forging for themselves, discussed elsewhere (Cristina Cerulli 2012b). Jay, for instance is working with a network of collaborators on a number of projects, one of which is a development of his thesis project AWE .

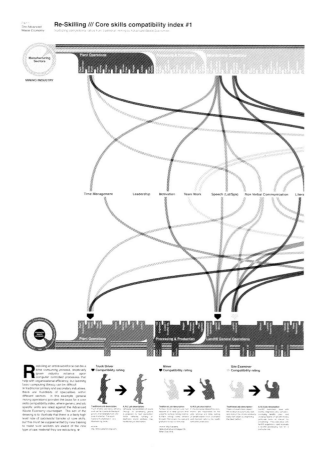

"In this example, general mining operations provides the basis for a core skills compatibility index, where generic and job specific skills are rated against the Advanced Waste Economy counterpart. The aim of the drawing is to illustrate that there is a fairly high level rate of successful transfer of core skills, but this must be supplemented by new training to make sure workers are aware of the new type of raw material they are extracting" (Jordan J. Lloyd 2009)

Fig. 3 Core Skills Compatibility Index

Part 2
The Procurement Process

The Advanced Waste Economy: A Visualisation
Agents and processes involved in the development of the Rother Valley

The Actor Cloud
A 'tag cloud' of stakeholders, influence represented by size of circle

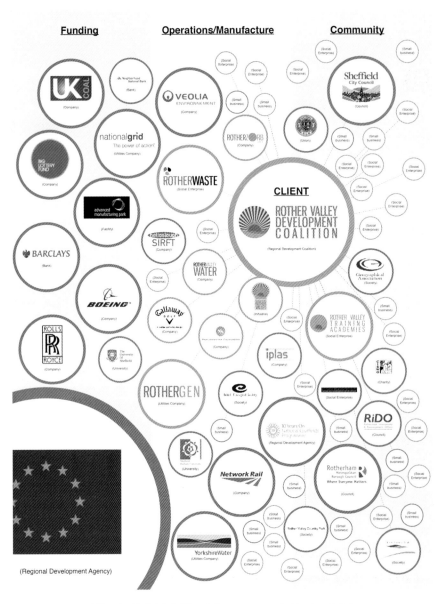

A 'tag cloud' of stakeholders, influence represented by size of circle

Fig. 4 Actor Cloud

Part 2
The Procurement Process

Advanced Waste Economy as a Project Finance Scheme

How an established procurement method distributes risks and profits between a number of enterprises (simplified schematic)

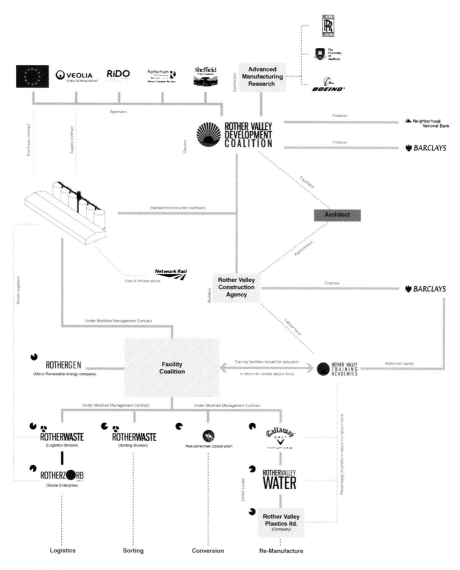

How an established procurement method distributes risks and profits between a number of enterprises (simplified schematic).

Fig. 5 Advanced Waste Economy as a Project Finance Scheme

三个作品、一个人物引发的建筑教育思考[1]
The Thinking of Architectural Education Aroused by Three Works and One Character

李国友　Li Guoyou
徐洪澎　Xu Hongpeng
哈尔滨工业大学建筑学院
School of Architecture, HIT

摘　要　"卓越工程师"是当下中国建筑教育界树立的一个新的培养目标。本文结合对学生的三个建筑设计作品和一位青年建筑师的讨论展开对理想化的"卓越工程师"的理想描述，提出"卓越工程师"需要具备的三种素质：独立思考的习惯和意识、职业道德和勇气、专业技能和智慧，并倡导将调整新的培养目标的过程当成建筑教育人文补偿的一次机会。

关键词　建筑教育，人文补偿，三个作品，一个人物，卓越工程师

Abstract　"Excellent engineer" is a new training objective established currently by the chinese communities of architectural education. Based on the discussion of students' three architectural design works and a young architect, this paper expand the ideal description of the ideal "Excellent engineer", raised "Excellence engineer" need to have three qualities: independent thinking habits and awareness, vocational morality and courage, professional skill and wisdom, and advocate that taking the process of adjusting the new training objectives as one opportunity of the cultural compensation in Architectural education.

Keywords　Architectural Education, Cultural Compensation, Three Works, One Character, Excellent Engineer

　　近来，"卓越工程师"成了中国建筑教育词库中的一个新宠，它似乎代言了未来一段时间内中国建筑教育的培养目标。如果按照理想化的愿望和标准来猜测"卓越工程师计划"，我们或许会形成这样的理解："卓越"，首先是指健全的心智和卓越的人格，充满职业公德心和人文关怀精神；而作为技术专家的"工程师"，则是指具备求真务实的科学精神的人。

　　建筑设计在许多场合被称作"创作"，大概是寄托了人们的一种期望，即让工程建造技术更大程度地兼具艺术色彩。继之而来，有了"创作思维"、"创作思考"，这使得许多人印象中原本让外行人艳羡的"无中生有"的"黑箱"过程变得更"黑"：有了创作概念撑腰，设计师可以超限度地表现个性，做自己思想的发言人，做新形式的缔造者，独揽新概念的解释权并为不断出现的不合理性进行专业辩驳。其实，在比较实际的意义上，设计就是用专业手段解决问题。这不是说把设计降低到了冷冰冰的技术问题层面，恰恰相反，"解决问题"本身就让人心里备感温暖，它是充满了人文关怀色彩的，尤其当解决的是需要解决的问题而不是建筑师所臆想出来的问题时。由于解决问题，建筑师才开始注意弱势群体并尝试将自己所学的本事用于关心他们的努力；由于要解决问题，建筑师才会对工程质量、技术和资金投入严肃对待，慎重下笔；由于要解决问题，建筑师才会克制自娱自乐的形式游戏和自我陶醉的无限虚荣，而真正关心起城市的历史和子孙的记忆。

　　在眼下我们的建筑设计和建筑教育圈子里，最难做到的大概不是技术的突破、风格的创新和形象的震撼力，而是某种"一致性"，比如：功能与功能类型表情的一致性；技术投入与节省原则之间的一致性；形式和空间的一致性；表皮和内容之间的一致性；投标方案的承诺和施工图之间的一致性……虽然我们在从事这个职业的过程中见惯了种种不一致，但我们对这些现象的麻木则是更大的危机。这种危机表现为一种沦落了的职业道德，并随之加速地降低设计师的鉴赏力、设计水平并最终恶化我们的生存环境。

　　对于这种状况，建筑教育需要承担责任。改变的一部分希望自然

1　本文为"2011中英韩建筑教育国际论坛"宣读论文。

寄托在未来的建筑师身上。按照理想化的标准,卓越的设计师大概需要具备以下三种素质:独立思考的习惯和意识;职业道德和勇气;专业技能和智慧。虽然专业现实的种种不良走向在逐渐侵蚀教学环境,但是年轻的学生仍然饱有让人感动的职业智慧、勇气和纯洁的品德。让我们来看一看三个出自建筑学专业在校学生的作品,这是三个虚拟的构想。

方案一:青年寄宿社区

这是2010年中联杯大学生建筑设计竞赛的参赛方案。任务书提出的要求是为一些年轻而经济上相对清贫的创业者提供某种可能的居住和交往空间。任务书在解释题目时还举了上海的例子,描绘了类似于"北漂"一族的青年创业者的困境和缺少社会关注的现实问题,提出了为这一人群打造低廉、活跃、方便的生活条件的理想化模式。参赛作品的可贵之处是不仅避免了陷入形式和空间游戏的"创新"误区,而且敏感地将任务书所描述的现实问题通过"借力"的方式予以展开和转述,达到了一种职业的敏感和思考的深度。虽然众多业外人士质疑这座天价大楼的外观形象,但是,从专业角度看,更需要质疑的是大楼平面布局所存在的不合理关系和结构系统潜在的安全隐患。事实上,一位政协委员连续十几次提出改建CCTV的建议:根据他的研究,一旦与地面直接联系的两条大腿的双柱体部分有一个柱体发生火灾、竖向交通难以发挥作用时,所有的人群都将瞬间汇集到向另一条竖向交通核大腿上。除了那条烧着的大腿弯曲所带来的L形悬空部分整体倾覆的危险外,庞大人流的瞬间集中将导致灾难性的交通混乱。学生们的设计概念非常简单,用CCTV过于充裕的钱打造空中青年社区以周济缺少帮助的年轻人,同时在技术上解决结构平衡和交通疏散的危机,在态度上弥合CCTV实际行动与宣传理念之间的鸿沟。如此,两全其美,在救济别人的过程中完成了对自己灵魂的救赎!《金陵十三钗》的精神线索在这里以更加合情合理的方式得到了再现(图1)。

方案二:都市新型信访接待中心

这个作品同样是投送中联杯大学生建筑设计竞赛的一个设计方案。这一年的设计竞赛题目是"城市客厅",单从这四个字本身,我们就足以体味出题人的美好愿望和人文情怀。在梳理了各种可能的思路和切

图1 青年寄宿社区,2010中联杯大学生建筑设计竞赛参赛作品哈尔滨工业大学王一申、王墨晗、于洪磊,指导教师李国友

入方向后，一个大胆的思路渐渐从众多想法中显露出来。学生的出发点非常好：最温暖的城市客厅，给最需要关心的城市人群。确定这一点之后的关键一步是寻找最需要帮助的城市人群，学生的目光很快锁定在备尝艰辛的"访民"——为了维权而上访的城市居民身上。考虑到这一人群极度渴望社会，尤其是政府相关部门的关心和帮助，因此访民的城市客厅似乎需要处于视野开阔的城市中心，而这一人群又常常伴随着激动的情绪和行为，因此提供足够的空间和必要的公共设施条件以疏导和缓解他们的负面情绪就变得极为重要。最终，学生们将地段选在了某北方城市中心区三条道路交汇的一块三角形开放休闲广场上，设计中运用下沉、抬升等多种办法创造了相对开阔的空间和足够面积的停留、休息、交流、表达场所。大概是由于最终的设计成果过于简略、没有很好地体现设计概念的缘故，作品最后没有获奖，但是，学生大胆运用所学知识为城市弱势群体提供关怀和帮助的勇气却给我们这些习惯在世俗禁忌下思考的头脑吹来一股清新的空气（图2）。

方案三：某可持续建筑设计竞赛参赛作品（图3）

这是个用"解释"代替"设计"的设计作品，以"转译"的方式表达对当下建筑设计界时弊的质疑，具有一定嘲弄和调侃的意味。作品正面回应设计题目"可持续建筑"，设计答案就是一个简单的动作——遍贴"可持续"标签！在这里，设计的对象不再是建筑，而是一种包装方式。正因如此，具体的建筑形象和内容不再是作者关注的对象，除了统一的标签"可持续"字样或口号之外，建筑可以随手拿来（或称"借用"），无需专门设计。作品的巧妙之处在于："随手拿来"这种随机动作使得所有的建筑都有了华丽转身为"绿色建筑"的可能，这种"全集合"特征再现了当下不少挂着标签的所谓"可持续建筑"其实与原来的建筑没有区别的窘迫真相！作品将虚假的"可持续"冠名撕下，并将其痛痛快快地嘲笑了一通。作品的珍贵之处是一种独立思考、发现问题的专业敏锐和当众嘲讽皇帝新装的勇气。

发现这三个方案的过程很偶然。我在邀请一位专业课非常优秀的学生参与一次关于竞赛的讨论会时，他向参会者推荐了第三个方案，他认为这个方案发人深省；在参与组织学生集体参加设计竞赛时听到了第二个设计，当时我的反应是：好想法，视角独特，真正从人文关怀的角度发现问题并不容易。我在指导设计竞赛的过程中与学生一起激发出第一个想法，既然做了，就做一个有创意、有启发的实在

图2 都市新型信访接待中心，2010中联杯大学生建筑设计竞赛参赛作品 哈尔滨工业大学建筑学院 张立、刘峰，指导教师 邹广天

方案。我常常想，这三个方案大概代表了目前建筑设计专业的学生思想中一些有希望的东西。对照我在前文所述"卓越的设计师"应该具备的"三种素质"：第三个设计的希望在于其"独立思考和质疑的意识"；第二个设计的希望在于"职业道德和勇气"，即专业服务对象的平民化；第一个设计的希望在于这些未来的建筑师们不但在积累"基本技能"，还在逐渐积累"专业智慧"。

最近，震动世界华人建筑圈的最大事件是普利茨克建筑奖颁发给了一位特立独行的年轻建筑师和教育工作者——王澍。抛开面子上的事不说，这件事的意义大概更多的在于一种潜在的"示范"效应：追赶世界潮流者被潮流甩远，有独立思考的本土文化实践者却一跃走到潮流的前头！王澍的特立独行体现为他完整地保留了似乎只有学生才有的一种理想主义情结、相对要少得多的功利思考和平民化、本土化的人文关怀精神。这些，被他的独立思考习惯和个人艺术素养、专业智慧、实践经验乃至个性语言烘托得非常极致。从这意义上讲，王澍倒可以算得上是一位"卓越"的"工程师"。

摘两段王澍的话，我们会有一个直观的印象："中国建筑现在的发展状态是非常畸形和混乱的。一是畸形繁荣，因为数量巨大而陶醉在一片欣欣向荣之中；一是畸形的专业化，在非常大量而且非常快速地建造的情况下，建筑师基本上没有什么时间思考，甚至包括专业上很多基本的问题。""现在的建筑设计专业总是在几个狭隘的概念里兜圈子。很多本科生一来就学'某种'专业建筑学，老师也是这样教的，但问题是这个专业建筑学甚至对生活里发生的问题基本不关心。

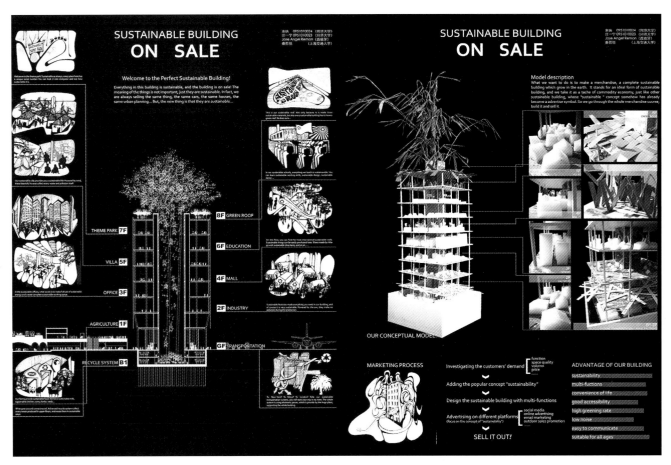

图3 SUSTAINABLE BUILDING ON SAIL 某竞赛参赛作品
同济大学建筑与城规学院：余快、汪一宁、Jose Angel Remon（西班牙）；上海交通大学：秦哲悦（指导教师不详）

所以我教学生的第一堂课都是'回忆课',回忆你的生活、你的家,你会发现非常有趣的是,每个人的生活差异是如此之大。例如,一般的住宅都是几室几厅,但我有个女学生却告诉我,她根本不愿意住在自己的卧室里,而是今天跟妈妈睡,明天跟姐姐睡,她这种生活方式就是很特殊的。""当建筑和基本的生活失去了联系,而只是按照一些专业的常规在给别人安排生活的时候,这个专业就会非常狭隘。"在设计杭州钱塘江边的"垂直院宅"时,他说"我试图恢复对普通人来说有意义的生活,而不是对一个艺术家来说有意义的生活。"

需要补充的是,思考和思考习惯是完全不同的两个事物。对于任何一个个体来说,在某一特定时期都会出现一些独立思考的现象,但这并不能保证有持续的结果。毕竟,负责任的理想化批判和不负责任的现实做法之间的"毫不相干"和"自成体系"往往会达成一种奇怪的平衡,这种奇怪的平衡甚至出现在一个人或一个团队的身上,这也实在堪称中国大陆建筑设计和建筑教育界的一大奇观。

人有某种虚荣倾向,正如"博士学位"近乎泛滥并几乎沦为基础教育标准一样,"明星建筑师"的含金量也已经大大贬值。今天在中国的建筑设计圈子里,明星建筑师并不一定是"卓越工程师",至少从一个更独立、更负责任的意义上看是这样。有多少知名建筑不是靠着某种"高碳"投入所产生的震撼效果和建筑摄影家穷其所能拍出的极具煽情效果的图片而获得"荣耀"的?作为建筑最基本属性的效能及真正使用者的评价早已无人过问了。这不能不说是当下建筑教育"工具理性"教育逻辑的结果。因此,当我们面对"卓越工程师"这一宏伟"计划"时,我们必须充分警惕其中潜在的危险走向,并努力使之校正到一个更为健康的方向上去。

(资料来源百度空间•vivi记事簿:http://hi.baidu.com/ava_vivian/blog/item/44da0cd87329be3633fa1c8f.html)

Embracing the Positive Grey: Revealing Architectural Ideas[1]
拥抱积极的灰色：揭示建筑理念

Satwinder Samra
School of Architecture, University of Sheffield, Sheffield S10 2TN, United Kingdom
谢菲尔德大学建筑学院，谢菲尔德 S10 2TN，英国

Abstract This short essay is based on a lecture that was delivered at the 2011 China-UK-Korea Trilateral Conference on Design Education at the Harbin Institute of Technology. Positive Grey is a terminology I often use to describe and highlight the portion of the design process that exists part way through a design process. All designers acknowledge the importance of creativity during the inception, development and resolution of a project. Design is a fascinating, rewarding and at times arduous process that often requires great leaps of faith, courage and intuition. With personal hindsight as a practitioner and the experience gained from teaching multiple cohorts of architecture students in many institutions, I have realized that this is not an uncommon experience. This paper highlights some of the tactics that I encourage students to adopt when navigating and developing their work. Topics discussed include The Dilemma of Time, The Perils of Originality and The Value of the Student Project.
Keywords   Ideas, Creativity, Architecture, Students, Context

摘　要　本文是基于2011年在哈尔滨工业大学主办的中英韩建筑教育学术研讨会的演讲而成，积极的灰色是我经常强调的一个术语，它是设计工程的一部分。所有的设计者都承认在设计的开始、发展和成果阶段创造性的重要性。设计是一个迷人的、有回报的同时也是艰难的过程，它依赖于我们的信心、勇气和直觉。教授学生的相关经验使我们认识到这不是一个普通的过程。本文强调的是作者在鼓励学生去适应、引导和发展他们的设计时所采用的方法与策略。本文将从时间的困惑，独创性的危险和学生方案的价值来展开论述。
关键词　想法，原创性，建筑，学生，文脉

Design is a fascinating, rewarding and at times arduous process that often requires great leaps of faith, courage and intuition. Through my own experiences as a practicing architect (with Urban Splash Liverpool, Proctor and Matthews Architects London and in my own practice Sauce Architecture with Daniel Jary) and as an Educator at Sheffield University School of Architecture[1] I have developed a number of techniques and tactics that allow me to negotiate and develop ideas at conception, development and resolution stages of a project. These have informed my own teaching and practice in many situations and scenarios. This short essay is based on a lecture that was delivered at the 2011 China-UK-Korea Trilateral Conference on Design Education at the Harbin Institute of Technology.

Reflecting on my early years as a student of Architecture, I remember that the studio based design projects often seemed confusing, overwhelming and difficult. With personal hindsight and the experience gained from teaching multiple cohorts of students in many institutions, I have realized that this is not an uncommon experience. This paper highlights some of the tactics that I encourage students to adopt when navigating

---

1　本文为"2011中英韩建筑教育国际论坛"宣读论文。

and developing their own work.

At the start of a project many students can be overwhelmed by the enormity of the task that lies ahead of them. To relieve this paralysis I have defined two key terms, which help to start to demystify the design process and evolution of ideas. These are 'Spinning Plates that Never Drop' and 'The Positive Grey'.

All projects require multiple agendas to be explored, revealed, developed and addressed simultaneously. The Spinning Plates analogy illustrates this notion. The designer needs to choose the plates; some will require more attention than others and this will change over the duration of the project. In order for the outcome to be successful certain plates may slow down but none should ever be allowed to 'drop' (Fig. 1).

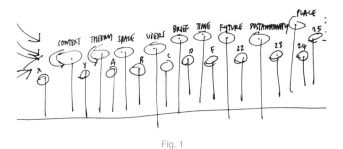

Fig. 1

I was really excited about the studio project, we had 6 weeks in total, the first was a mixture of research and thinking, then I got stuck, 4 weeks went by where nothing seemed to make sense, with little time left I had no choice but to make a model and some drawings of an under resolved project based on an unclear set of thoughts, the output was not well received and I was disappointed with my feedback and results! [2]

Positive Grey is a terminology I often use to describe and highlight the portion of the design process that exists part way through the project. All designers acknowledge the importance of creativity during the inception, development and resolution of a project. Design often has periods of doubt and difficulty. This often occurs in the middle of a project. This middle I refer to as the Positive Grey. This is something that students need to acknowledge, accept and embrace. By being comfortable with uncertainty and doubt they will not only enjoy the project more but also produce better, richer and more diverse work.

The Dilemma of Time and How to Exploit It

It is natural to think about ideas in the hope that the more thinking one does the better the ideas will be. I would suggest that this is a place of safety that can only encourage procrastination. This will not make the most of the time available. This could be called the 'Dilemma of Time'. To move forward ideas must be committed to paper (paper can mean any form of representation be it models, sketches, poems or collages) (Fig. 2+3). The very act of making and/or drawing allows the potential of a project to be revealed. The potential

Fig. 2

Fig. 3

Fig. 4

of the project is always waiting to be triggered; intuition will not make this happen but action will.

'For me (drawing) can be a way of exploring a preoccupation (sometimes leading to the discovery of another). At other times it is an intense relationship with the actual materials of the drawing, and the moment, which brings it into existence. Drawing can be a way of overcoming fear. At different times it is all these things.' [3]

The methods adopted should be multiple, sometimes slow, more often quick and ideally communicate directly either to yourself or others. The rise of multiple techniques of production in recent years could be viewed as making the process easier. However using tools without judgment or direction can be debilitating in itself. Using technology too soon can distance the designer from their ability to see, perceive, visualize and represent that moment in time. We often encourage freehand methods at all stages of the process. The connection between eye, mind, hand and pen/pencil and paper is vital. This is not to discourage the use of computers and other digital tools but to see these as part of a repertoire of available techniques (Fig. 4).

Looking for Opportunities Not Answers

Students often arrive at the School of Architecture with an accepted model of learning which assumes that if a set body of knowledge is consumed it can then be regurgitated when the need arises. This is not only constraining, but also an obstacle for the designer. I would suggest that drawn speculations will lead to opportunities, which will in turn lead to propositions that are based on investigation of the situation itself rather than a predetermined set piece answer. One only needs to look at the Modern Movement in all it's guises to see an oven ready mantra re-applied verbatim across the globe, without an understanding of the local or specific conditions. It is no wonder that the users of these environments feel let down and isolated by their surroundings.

Include the Viewer and the User. Exclude at Your Peril

Users/ Clients/Stakeholders often do not see space and its subsequent use as an abstract construct but through their everyday lived experiences. The challenge for the designer is to find meaning and mechanisms to allow these to be revealed. We often encourage drawings that exploit and reinforce visual codes to allow people to be brought on board using typologies that exploit visual readings through the use of diagrams encouraging visual dexterity and tactical representation (Fig. 5+6).

Listening Designers and the Legacy of the 'Crit'

The Architectural profession is often maligned for its inability to take on boards the views of others. This I would suggest is in part is due to the Architectural language that is used. This can be inaccessible, obscure and elitist. The other factor that can posit a defensive attitude on the part of the designer lies in the legacy of the educational platform of the crit. Here students' present work to visiting critics who need to be questioning and inquisitive but can often be overbearing and aggressive in their review of the work presented. It is no surprise then that the student adopts a defensive approach in an attempt to 'protect' their work.

This approach is often taken into their professional lives

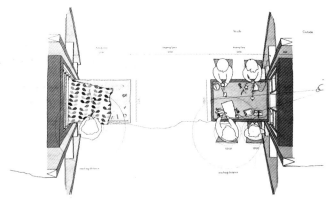

Fig. 6

Fig. 5

where Architects promote their own ideas rather than take on board the views of others. A number of informed and thoughtful studies have taken place here at the University of Sheffield about the place of the 'Crit' and it's impact on subsequent generations of Architects[4]. Our 'Crits' are now called reviews where the space of the presentation is seen as a place of collaboration between student, reviewer and peers. As opposed to the pageant of eminent critics mutually nourishing their own ego using the student work as a backdrop. Obviously the student needs to develop the ability to advocate and promote their work but this should be with support and direction. The reviewer has a responsibility to act as an educator in the process. At Sheffield University School of Architecture we actively encourage staff to take a more forward thinking approach as educators (Fig. 7). In doing so I believe we may foster a generation of less arrogant, more open and inclusive designers who may be able to respond to the challenges of our time.

Context

Projects do not exist in a vacuum- often the computer screen or the A4 piece of paper can become the soulless context for ideas to exist. The context (real, imagined or otherwise) should be ever present. Time taken to investigate the physical, economic, social and political environments will always pay dividends. 'Thinking about space one needs to appreciate the cultural context otherwise one just gets shelter or a useful object. There needs to be constant mediation between the situation and the context that exists at that time.' [5]

By doing so we can avoid the classic scenario where context is only brought to the project at the end, often with disastrous consequences (Fig. 8). Many architectural projects have been discounted not only in Architecture schools but also during prestigious architectural competitions, where the designer has ignored or misinterpreted the existing context and fabric. This lack of respect is often exemplified where designers not only fail to acknowledge the existing condition but also fail to draw it!

Revealing the Potential of the Project

The onus for this lies with the designer. By adopting an open

Fig. 7

Fig. 8

Fig. 9

Fig. 10

and thoughtful approach it is possible to speculate with direction. Adopting a multi pronged approach can do this. Varied methods of working can allow the designer to reveal and see what is possible. This is encouraged in my work at the University of Sheffield using Project Swap Sessions where students work on each other's work through live drawing and dialogue (Fig. 9+10). Alvaro Siza suggests that 'making sketches without making at the same time more rigorous drawings is good for nothing. It is absolutely necessary to do both at the same time.'[6] He goes onto advocate in his teaching that a range of drawings must be used to explain a project. It is refreshing to see an architect being explicit about his approach: 'a choice of drawings showing every moment… first to demystify the project and show that it is something always very systematic and coming from a certain knowledge.' [7]

Often many designers are not open or candid about their processes of approach to revealing and developing ideas- they prefer to hide behind the notion of Genius. Siza admits that although they are often poor ideas, experimentation is fundamental. The ideas must be committed to paper as they come, so they can be pursued or discarded[8].

There are numerous theories, tactics and initiatives that designers adopt whilst developing a project. Many keep this undisclosed perhaps as a way of perpetuating the 'myth of genius'.

Through my teaching I try to uncover and dispel this myth. Obviously certain students have a natural talent and pick things up quicker but others just need to practice and more importantly produce. Perpetuating the Myth of the genius is not a fruitful or useful act as it continues to promote architecture as a heroic individual act as opposed to a thoughtful, investigative reflective yet collaborative mission.

The Value of the Student Project

This leads me onto the true value of the student project and it's importance as a place to open up, reflect and pursue thoughts, ideas and speculations. The student project by it's vary nature is theoretical. This allows one to speculate without having to build; in certain circles this is seen as irrelevant or self-indulgent. The discourse one can achieve through drawing and model making alone is a powerful and at times liberating exercise (Fig. 11). This should be celebrated 'One of the powers of Architecture is it's non-dependence on realization. Drawings can dismantle or disregard material and weight providing insight into the unobtainable.' [9]

Nothing is Original: Everbody is Influenced by Something

Being provocative I would suggest and often have done that nothing is original in Architecture (and many other creative disciplines). We are all influenced by culture, background, perception, insight, opinions and factors of our previous and current environments. This should be celebrated and brought into the work. For an architect to understand and design for others they must first acknowledge their own lived world experiences, this without doubt holds the most potential. 'You can never claim to understand culture all you can do is live in it.' [10]

We can be encouraged and take note of creative individuals from other disciplines such as Paul Arden who was creative director at Saatchi and he insists that we should 'Devour

Fig. 11

films, music, poems, photographs, dreams, trees, architecture, street signs, clouds, lights and shadows' [11].

If one were to select a number of journals over a 50 year period it would not take long to realize that all architects are influenced, steered, directed by the body of knowledge and output that has gone before them. Originality and genius are myths perpetuated by the publicity loop of media hungry Architects and Journalists keen to celebrate the spectacle of the 'new'. It is important for students to realize that the quest for originality will only hold them back. They should embrace being influenced, steered by that what has gone before them and in doing so will reap the benefits of the wealth of human experience and endeavor that has shaped the world we live in. This brings us onto ideas. 'if an idea is not taken up and used as a solution to a problem it has no value.'[12] This could be taken further by suggesting that if ideas that are not revealed or tested, they do not actually exist. By seizing the potential of an idea or notion through drawing it in real time, it will become nourished and nurtured. Often students suggest that they can see in an idea in their mind. I would suggest that it couldn't exist until it is represented, developed and more importantly shared. This brings us to a fundamental aspect of design development, i.e. the way forward often comes from conversations with others not from an isolated hermetic individual sat alone in cyber space. I would advocate that the physical design studio should be promoted as a place of conversation, collaboration and a place for design to nourish.

In conclusion I would suggest that students who are canny, flexible, reflective in their approach and working methods could produce rich, vibrant and inclusive architecture. By embracing the Positive Grey and adopting some of the tactics suggested they could reveal the true potential of their creative endeavors.

Notes:

[1] The Author has extensive teaching experience in a number of institutions. He was Head of the Year 3 Undergraduate Program from 2000-2010 at Sheffield University and has also taught and Reviewed at Manchester SOA, Leeds Met SOA, Mackintosh Glasgow, Leicester SOA, RCA London and The Bergen SOA Norway.

[2] Anon Student Feedback circa 2002

[3] Alison Wilding quoted in Leeds Sculpture Collections. Works on Paper

[4] See 'The Crit' by Rosie Parnell, Rachel Sara, Charles Doidge and Mark Parsons for a detailed discussion on this mode of Architectural Education.

[5] Author quoted in Interview with Pierre D'Avoine and Simon Bowen in 'De Re Metallica' unpublished DipArch Dissertation. University of Sheffield 1993

[6] Alvaro Siza quoted in ' Why Architects Draw' Edward Robbins : MIT 1997 p154

[7] ibid p157

[8] Alvaro Siza quoted in Richard Marks Lecture Review 'Architects Journal' 12th March 2009

[9] Mayne, Thom 'Connected Isolation' Lecture published in Architecture in Transition edited by P Noever p79. Munich, Prestel 1991

[10] ibid

[11] Paul Arden 'What ever you Think, Think the Opposite' Penguin 2006 p94

[12] ibid p94

讲述
Illustration

教学讲座

The Teaching Lecture

Establishing University Based Activist Research —— the Portland Works Project[1]

基于大学的推动研究——以波特兰项目工程为例

Cristina Cerulli
School of Architecture, University of Sheffield
谢菲尔德大学建筑学院

This talk will reflect on the aspects of activist research and collective production and action to create a shift towards more sustainable business, cultural and civic communities. This reflection is situated within the context of the Re-imagining Portland Works Project, a Knowledge Transfer project concerned with helping the local community to imagine a future that is environmentally, socially and economically sustainable for Portland Works, one of the outstanding examples of Sheffield's industrial heritage. In particular the complexities and the mutual interplay of actors and networks will be discussed in the context of a collective production that encompasses the cultural, business and civic spheres. Furthermore the strategic role that academia can play, through research and research led teaching, in facilitating the emergence and successful establishment of collectively powered social innovation will be discussed.

The Portland Works campaign started out in January 2009 in opposition to a planning application to convert the works into small bedsit flats. This redevelopment would be the end for many of the businesses based at the works that could not afford to relocate or could not find suitable alternative accommodation.

"The Portland Works Campaign has two aims: Firstly, to oppose the planning application currently submitted by the owners, which would effectively close the existing businesses and convert the building to residential use. 66 studio flats are proposed.
Secondly, to seek ways to secure a long term future for this fine building and its unique collection of businesses and creative people."

The involvement in the campaign crosses several of these projects, but the focus of this talk is in particular: the Reimagining Portland Works: Sustainable Futures for Sheffield's Industrial Heritage Project. It Funded by the University of Sheffield Knowledge Transfer Programme, our

---

1 本文根据Cristina Cerulli于2011年9月8日在哈尔滨工业大学所做的讲座整理而成。

SHAW ENGRAVING

ARTIST MARY SEWELL

BAND - THE GENTLEMEN

PORTLAND ELECTRICAL

PAUL - QUALITY CABINETRY

SEQUOIA SOUND STUDIO

STUART MITCHELL KNIVES

MARK JACKSON - SQUAREPEGS

PETE LEDGER - PML SILVER PLATING

PAM HAGUE - PH ENGINEERING

ARTIST CLARE HUGHES

LYNTHORPE WOODWORKS

project was concerned **with helping a local community of makers, small businesses and creative industries to imagine a future that is environmentally, socially and economically sustainable.** Through research and activist civic action we have sought to investigate the collective knowledge-production of strategies, tools and tactics available to **economically threatened communities** to enable the envisioning and enacting of sustainable futures. The approach was to develop **and implement a framework for collective production and action** where engaged scholarship, community activism and community economic development converged to actually save Portland Works from speculative redevelopment and to retain it as a place of making. In doing this we were addressing questions of how businesses can become more resilient to market pressures from another sector (in the case housing), how agency could be created within networks of tenants and how new productive relationships could be formed that would contribute to the sustainability of Portland Works.

Researching Precedents

As part of the Knowledge Transfer Project 'Re-imagining Portland Works: Sustainable Futures for Sheffield's Industrial Heritage' a number of case studies were developed. The case studies were chosen to assist with the creation of a shared vision, governance model and appropriate business plan; **each represented models of management, ownership and funding or had an interesting driving ethos.**

Ten projects including Bank Street Arts, Stag Works, The Riverside, S1 Artspace, High Green Development Trust and Butcher Works in Sheffield, Coexist/ Hamilton House in Bristol and The Woodmill in London were used as the basis for one of the sessions for the Exploring Futures workshop, held in June 2010. The questions for the case study partners were devised in part through reviewing material form minutes of meetings with various partners, tenants and campaigners; this allowed us to compile a list appropriate to the campaign.

Each case study contained an introduction, location map, address and directory, an indication of the project's size and scope, a section on the ethos (people and goals), one on accommodation (occupancy and tenancy) and one on management ; each case study also included a chronological financial account of projects and a section on difficulties, disputes and failures. The information was gathered through interviews with a number of actors and through additional desk based research and site visits where required.

Through research and activist civic action we have sought to investigate the collective knowledge-production of strategies, tools and tactics available to **economically threatened communities** to enable the envisioning and enacting of sustainable futures. Our approach was to develop **and implement a framework for collective production and action** where engaged scholarship, community activism and community economic development converged to actually save Portland Works from speculative redevelopment and to retain it as a place of making. In doing this we were addressing questions of how businesses can become more resilient to market pressures from another sector (in the case housing), how agency could be created within networks of tenants and how new productive relationships could be formed that would contribute to the sustainability of Portland Works.

Recognition

This valorisation of 'multiple authorship' is it at odds with some reward mechanisms and values in academia, which privileges the single or lead author, who obtains knowledge from the object of their research. Rigour in activist research is about it being appropriate to its aims. Methodological rigour in activist research is of paramount importance as the lack of it could damage the very cause that the research is working towards supporting. There is a mainstream association of methodological rigour with the absolute control over the research process by the scholar; this misconception is often a barrier in recognising rigour within activist research, where

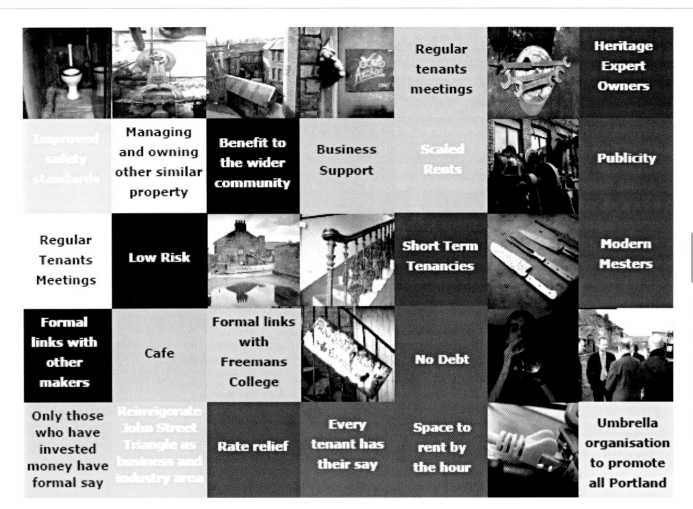

the commitment to collective and egalitarian knowledge production demand precisely the opposite: letting go of that control and engage in a research process that is open, responsive and horizontal. Activist research methods have also the built in test of validity of whether they are meaningful or they work for the participants that helped to formulate the research goals to start with.

Conclusions

The Portland Works Knowledge Transfer Project has investigated structures and approaches available to the campaign that can lead to sustainable, just and equitable futures for the Works. This has been a collaborative project, that has actually created empowering strategies and activities that are leading to the purchase and refurbishment of the building and the development of a viable business structure for its future management. Its strength as an approach came from its hybrid nature; it combined research with activism, and civic action with a political struggle for change. This contributed to aspects of the development, growth and sustainability of the project because it created a strong social group and wider set of networks, whilst also allowing us to develop shared values and vision that had a close relationship to business models and structures of ownership. By growing together both socially and politically, Portland Works is therefore stronger as a business. The ethical nature of the project makes it more sustainable because it is closer tied to what people want to offer and contribute and is relevant to their needs and desires. The multiplicity of approaches allows us to reach out to and become part of new networks and realises new opportunities and ways of doing with only limited resources. The openness and interdisciplinary nature of this approach allows us to innovate, drawing on a wider range of skills. Our awareness of the dynamics of collective civic production and action can enable and support social innovation and resilience by strengthening the signal of bottom up initiatives, creating significant changes and results even in the context of general scarcity of resources. In order for this project to happen we drew on resources from Sheffield University and Sharrow Community Forum, both financial and perhaps even more crucially, the well developed social networks, teaching and research support. We are making the case for intermediary organisations that can facilitate and enable this type of hybridism; it is crucial that policy recognises this to enable strategic and informed allocation of support.

Digital Ornament[1]
数字化装饰

Mark Meagher
School of Architecture, University of Sheffield
谢菲尔德大学建筑学院

The topic I would like to address in this talk is digital design, and the particular affinity between digital design methods and the tradition of architectural ornament. What I hope to show is that at least part of the current fascination with digital ornament is the result of an engagement with a search for the characteristics and possibilities of digital design. I will address three ways in which this connection between ornament and digital design has been described in theory and in practice: first, in terms of the reproduction of craft-based techniques; second, in the production of patterns; and third, in the representation of information.

If, as I believe is the case, Computer Aided Design has failed to deliver on many of its promises, why do we use the computer as a tool for design? What are those particular capabilities which it affords to us? What is the problem or need, to which the incontrovertible answer is digital design?

It's clear now that rather than transforming everything about design, digital tools have the potential to affect certain key aspects of the design process. Identifying these aspects with some precision is one of the key challenges for digital design in the coming years; and it's the topic that I'll address in the rest of this talk.

Before discussing in more detail the significance of architectural ornament, I'll take a moment to clarify what I'm referring to as digital design. For the pioneers in the field of computer-aided drafting and design, it wasn't at all clear what added value the computer might bring to the architectural design process, and many ideas were floated regarding capabilities that could entice architects to give up familiar methods of drafting and design in order to learn new, initially unwieldy tools.

The capacity for repetition has given rise to a counter-tendency for which digital design has been equally instrumental, the discovery of non-repetitive solutions that make use of the parallel processing power of the computer to produce repetition with variation based on rules. This development

---

1 本文根据Mark Meagher于2011年9月8日在哈尔滨工业大学所做的讲座整理而成。

is also anticipated by Sutherland in his identification of parametric design as one of the basic potentials of digital design which would shape its development in the future. He defines parametric design as the description of a given geometric entity as a set of relationships rather than a grouping of independent units. By defining a set of geometric relationships in Sketchpad, for example, it was then possible to run through the set of possible configurations allowable within the constraints defined in one's model.

Ever since John Ruskin's description of the craftsman in his book The Seven Lamps of Venice, there has been a persistent strand of criticism which faults the machine reproduction of ornament as a poor imitation of work produced by the human hand. According to Ruskin, the work of the craftsman possessed an individuality, and a quality which could not be imitated by the machine because this irregularity reflected the indecision and care of an individual person. The ornament produced by hand possessed originality, uniqueness, and an imprint of the author's humanity which could not be imitated by machine production. Ruskin also believed that while the building as a whole could come to reflect this intentionality of the craftsman, the hand-held object and the work of ornament were the most perfect embodiment of this ideal.

This ideal of beauty through irregularity has become a persistent aspect of the definition of architectural ornament. In the early days of digital design it seemed certain that the application of digital design and fabrication would succeed in solving this problem–the production of unique, original objects with the same cost and expenditure of effort as that required for serially-produced objects, an idea known as mass-customization, which can also be described as a customized product at the cost of the standard solution.

One way in which ornament has been used as a tool for describing the potential of digital design is in terms of a return to craft. It seemed to pioneers of numerically-controlled fabrication that the introduction of the computer in the fabrication process would allow the serial production of individualized products, an idea known as mass-customization. Through the kind of repetition with variation that we saw in digitally-derived pattern, numerically-controlled techniques promised the fabrication of architectural elements that are repeatable because based on a common rule, but unique because each is produced with different parameters. Just as artificial intelligence seemed on a near horizon, and it seemed that algorithmic logic would soon provide a means of imitating the subtle and unique variability of the human hand in its working of materials.

My point here is not to assess the success of digital design in imitating craft, but rather to ask whether the craft-based production of ornament offers a meaningful explanation of the potential of digital design. To answer this question, it's necessary to think more carefully about what it means to produce unique objects, whether by craft or using a robot following a sophisticated computer program.

Pattern is a type of ornament that has become particularly prevalent with the emphasis of many architectural projects on facade design. And, one of the most prolific manifestations of digital ornament is in the abundance of patterned surfaces of contemporary architecture. This type of ornament envisions a role for digital design as producer of surface finishes which make use of the parallel processing capabilities of the computer to simulate the result of many individual agents operating independently according to a simple set of rules.

I would like to suggest that on the whole, the analogy between ornament and digital design has been a positive one for the development of digital practices in architecture. The exploration of defining various types of ornament using algorithmic methods has led to a discovery of several previously unexploited possibilities for digital design, some of which I've highlighted in this talk.

Ornament is often the part of the building that exhibits the greatest level of visible complexity, and digital design is very good at producing complexity, or at least the appearance of complexity - so there's an inevitability to the adoption of ornament as a model for the integration of digitally-designed elements in the building.

At the same time, there is an alternate approach to the use of ornament as an aid in understanding the role of digital design. This is one that focuses on process rather than product - on how ornament was produced rather than the appearance of the outcome. This is an approach that is concerned primarily with the place of digital design in relation to the work of architecture, with the integration of digital design in the work of architecture. It is concerned with what tasks digital design is particularly good for.

In thinking about what is unique to digital design, it is important to consider the integration of digital design in the building. In other words, what is the question to which digital design is the answer? It clearly is not drafting. In the great majority of cases, it is not the design of the building as a whole. Perhaps one of the roles of digital design is to imagine the future of architectural ornament, one based in Variability and uniqueness.

The use of ornament as a means of describing the potential of digital design is as simple as the relation of a new technology to persistent topics in architecture. This connection is sometimes conscious, and often implicit. In both cases, the rich history of the term ornament has informed the implementation of digital design tools, and shaped the ways in which digital tools have been integrated into the design process. Although some aspects of the ornamental analogy have been misleading, on the whole the analogy has led to a more creative, and more integrated approach to digital design.

Urban Open Spaces:
Their Importance for Daily Life[1]
城市开放空间：城市生活的重要性

Helen Woolley
School of Architecture, University of Sheffield
谢菲尔德大学建筑学院

Introduction

The lecture is mainly about the importance of open spaces in our daily lives and is based on the book *URBAN OPEN SPACES*, which was written 9 years ago. One purpose for writing the book was to bring together evidence to show that open spaces have many benefits for our lives. In England, we have many cities which have parks and green spaces, and many of the parks were given by industrialists, who wanted to have green spaces in the city, so that people had opportunities to be relaxed and benefit for their health away from their work. During the 20th century, some of us in England forgot that these green spaces and parks are important: we put less money into them and we neglected them so some of them began to look very shabby. I partially believe that green spaces are important for our lives, so I wrote this book. If I was to rewrite this book, there would be a lot of more evidence to explore.

I want to address two questions: one is that what is open space, and the other is why is open space important in cities and urban areas?

What is open space?

Various people have tried to define what open space is. Some people have suggested that open space is the land and water in an urban area that is not covered by cars and buildings, or any undeveloped land (Gold, 1980). Someone else suggested that open spaces are not only the land, or the water on the land, but also the space and light above the land (Tankel, 1963). Someone in America suggested that open spaces are wide-open areas that can be fluid to the extent that the city can flow into the park and the park can flow into the city (Cranz, 1982).

Some people use the term 'public space' which can have different meanings for different people. Someone argues that public space is responsive, democratic and meaningful places that protect the rights of user groups (Carr, 1992), and I could

---

1 本文根据Helen Woolley于2012年9月5日在哈尔滨工业大学所做的讲座整理而成。

question that very strongly because some user groups are excluded from certain open spaces. Ali Madanipour (1999) said that public spaces are areas within towns, cities and countryside that are physically accessible to everyone, where strangers and citizens can enter with few restrictions. And someone said that public spaces are any place that people use when not at work or at home (Shonfield, 1998), and we can discuss this and may not agree with this because we may use public spaces in our lunch break at work. Walzer (1986) suggested that public space is space where we can share with strangers, people who aren't our relatives, friends or work associate. I might disagree with that part of this definition because sometimes we go with friends and relatives as well. This definition goes on to explain that public space can be space for politics, religion, commerce, sport, or for peaceful coexistence and impersonal encounter. Its character expressers and also conditions our public life, civic culture, everyday discourse. Another piece of work in England suggests that car boot sales and supermarket cafes are public spaces (Mean and Timms, 2005) and as a landscape architect I find this a strange type of public space and not an open space.

As professionals in England, sometimes we need guidance to help us work with the same understandings across our profession. After the great decline of finance to parks in England, there were a lot of policy making about our parks and green spaces between 2002 and 2010. And one thing that the government developed was to develop a typology of open spaces: defining what open spaces are in order to help professionals work to a shared understanding. The types of open spaces are: public parks and gardens, natural and semi-natural, green corridors, outdoor sports facilities, amenity green spaces, spaces for children and teenagers, allotments and urban farms, cemeteries and churchyards. These make up a green network of a city.

Defining urban open spaces from a user's point of view. When I was writing my book, I was thinking about the different types of open spaces not from the point of view of a planner as the government's typology has done, but I thought about it from the point of view of the users of the spaces. So I came up with a different typology, thinking about it as if we are at home, what open spaces might we use. So I identified three categories: Domestic open spaces, Neighborhood open spaces, and Civic open spaces.

To begin with, domestic open spaces include the area close to where someone lives, the housing area. Second, there might be private gardens where people live. Third, sometimes there are community gardens. These can take different forms, and the most famous is the lower east side of New York where people grow flowers, vegetables and food. In England, the most usual form of community garden is a shared garden where there are flats. Fourth, within the domestic category of open spaces are allotments.

For neighborhood open spaces, the most well-known open spaces are our parks. Parks are very important. In your park (when I visited it here in Harbin) I had to pay to get in, but in our Sheffield parks, we do not pay to get in them. Another neighborhood open space is playgrounds. Someone said that a playground should not be 'an island', but one of the places that children have access to in the neighborhood. The third neighbourhood open space is sports pitches: in England, we have lots of football pitches. And I am quite critical because football pitches are used for a very short time, perhaps 4 hours a week, by very few people, and we spend a lot of money cutting the grass meaning that they have very little biodiversity. The fourth neighbourhood open space is school playgrounds. They are used for play and they can also be used for educational opportunities. Sometimes people think that children do not play much on school playgrounds but one of my research projects identified a lot of different types of play taking place: play with high verbal content; play with high imaginative content; play with physical content; less structures play including sitting, etc. The fifth neighbourhood space is that of streets. In Sheffield, I love it when it snows because the street goes quiet and changes the whole atmosphere. A sixth neighbourhood open space in England is city farms. These started in 1972 in London and offer opportunities for children to have contact with nature: animals, not just plants. Sometimes there are some green open spaces that we cannot easily put into a category: for example, SLOAP (Space Left Over After Planning) and temporary sites. Sometimes the local government does not think these open spaces are important, but sometimes they can be very important to local people.

Regarding civic open spaces, first, we have commercial spaces. There is a statement that without the square there is no city and for many cities the square is its historic heart. Second, civic spaces can take the form of health spaces such as those associated with doctor's surgeries or hospitals. These are not new. It is just that as we went through, the 20th century, we forgot the importance of open spaces associated with hospitals and other health related buildings. Take the example of the university where I work: within half a mile, there are five or six hospitals and no open spaces associated with the hospital buildings at all, except one of the hospitals has two small courtyards. The third civic open space can be understood as those relating to transport. For example, historically many transport routes were for trade and movement of people. In time routes became roads and dirt became tarmac. Then canals and railways were developed. In England we had a period of industrial revolution when a lot a canals were built. The fourth civic open spaces can be identified as recreational spaces. This can include for example woodland, urban forests, golf courses and cemeteries. Another recreational use of civic open spaces can be skateboarders who like to use some of these spaces for their recreation.

In summary, concerning the hierarchy of open spaces, there are people who try to define whether open space is private or public, whether it is semi-private or semi-public. Another question is: who owns, uses, manages the open spaces? And these can have a great impact on all the types of open spaces.

Why is open space important in cities and urban areas?

So the second question that I want to address is: what are the benefits of open spaces? I will identify these benefits it into several types: social, health, environmental and economic.

Regarding social benefits, let us first consider children's play. We know from research that we have undertaken for our government that 'taking children to play in the local park is the most frequent social use of urban parks and playgrounds in England' (Dunnett, Swanwick and Woolley, 2002). A second benefit is that of the provision for active recreation. For a while, in England, there is a sort of cultural understanding that sports are very important in our research it showed that only 6 % percent of users of parks and urban green spaces go for sports. So active recreation is important, but the number of people who undertake it is not as high as people think. The third benefit is passive recreation. It is not about sports, but other things that can be done in parks. Passive recreation is the main use of urban parks in England, and more people go for passive activities than events and active activities. Fourthly, our open spaces can be a community focus, for instance for a local festival, school sports day or plant sales. A fifth benefit is that of providing a cultural focus. People from different cultural backgrounds sometimes use open spaces in different ways. But sometimes the cultural focus is actually a reflection in the design of the space. The sixth benefit of our open spaces is that they can be an educational resource. For example there was a rivers curriculum project in USA and in England we have had a national educational manager for British Waterways.

Concerning the health benefits: health is a statement of complete physical, mental and social well-being. Living in a green area is healthy for people. For the physical health, and we have an increasingly aging population, it is suggested in England that adults should have half an hour of physical activity five times a week. We have increasing levels of overweigh and obese children in many parts of the world. In England children are recommended to take one hour of activity each day. For mental health, benefits are: recovery form stress and restoration from fatigue. Even three minutes in the green space can help relax your brain.

One of the biggest issues that we have in cities is the climate of the city which is affected by all the glass, concrete and tarmac and results in what has been called the 'urban heat island effect' (Lowry). So the environmental benefit is how parks and green spaces in cities can affect airflow, air pollution, radiation, sunshine and precipitation such as rain, hail and snow. All these can be affected by the cities, and our parks and green spaces can help lessen the impact of these negative things. Moving on, the next environmental benefit is wildlife habitats: parks and green spaces in cities provide places for animals including insects, birds and mammals to live. Another of the environmental benefits is the fact that people can experience wildlife. Human contact with nature is very important because we are part of nature and parks and green spaces in cities provide opportunities for people to have contact with nature.

The next benefit of open spaces in cities is the economic benefit. First parks and green spaces in cities can have a positive impact on property values but this is complex and can depend upon the area, the types of park, the local population and the types of property. A second economic benefit is that of employment opportunities. For example, gardeners, rangers, street force, consultants, staff at city farms and landscape architects can all be employed working in and for and in our open spaces in cities. Thus many people benefit from our open spaces because they are employed in them. The third economic benefit of open spaces is tourism and in London this takes the form of Kew Gardens and Highgate cemetery, to name two specific open spaces that are visitor attractions.

References:

[1] Carr, S., Francis, M., Rivlin, R. and Stone, A. Public Space. Cambridge: Cambridge University Press, 1992.

[2] Cranz, G. The politics of park design: A history of urban parks in America. London: MIT Press, 1982.

[3] Dunnett, N., Swanwick, C. and Woolley, H. Improving Urban Parks, Play Areas and Green Spaces. London: Office of the Deputy Prime Minister, 2002.

[4] Gold, S.M. Recreation Planning and Development. New York: McGraw-Hill, 1980.

[5] Lowry, W. P. The Climate of Cities. Cities: Their Origin, Growth and Human Impact, Readings from Scientific American. San Francisco: W. H. Freeman and Company, 1967.

[6] Ali Madanipour. (1999) Design of urban space: An inquiry into a socio-spatial process. London and New York: Wiley Mean, M. and Tims, C., 1999; People Make Places:growing the public life of cities. London: Demos, 2005, supported by the Joseph Rowntree Foundation.

[7] Shonfield, K. The Richness of Cities: Urban Policy in a New Landscape. London: Comedia and Demos, 1998.

[8] Tankel, S. (1963) 'The importance of open spaces in the urban pattern', in L. Wing (ed.) Cities and spaces: the future use of urban spaces, Baltimore: Hopkins.

[9] Walzer, M. (1986) Public Space: Pleasures and Costs of Urbanity, Dissent 33(4): 470-475.

How to Be a Happy Architect?[1]
如何做一名快乐的建筑师？

Irena Bauman
School of Architecture, University of Sheffield
谢菲尔德大学建筑学院

What I want to talk to you about is the next 50 years, and the period in which all of you working as an architect. So I would like you to think about what it will be like in the future after you finish your studying here.

So to start with, I would like to introduce myself.

This is the office of my work, and I built it about 5 years ago. We bought some land in a very poor neighborhood in my city of Leeds in England. And this is my team, who want to say hello to you. Although the team changes frequently, the atmosphere is always happy, interesting and hardworking. We always have discussions, because we are interested in what the future holds for us. We find it necessary tore fresh our knowledge all the time. This is Sheffield University, and all of our students work in the top floors of the building. These are my master students, and we are always traveling together, because the best way to learn is to use our eyes.

But most students and architects asked themselves the same questions over and over again :what are we get out of bed for in the morning? What is worth while doing each day?

We think it's about doing something that makes you happy. So now I will run through some of the drivers of change and some of biggest challenges that will influence the work you students wherever you are and this will allow you to decide how you can respond to these and whether this will make you happy .

The global population is going to exceed 9 billion after 30 years, and most of the new population will be poor. So we are trying to build homes for the poor people, but we can't do it fast enough and we don't have time to plan proper neighborhoods with all the facilities that a community needs. So we build skyscrapers to fit the growth of population but not the social infrastructure that needs to go with it.

1　本文根据Irena Bauman于2012年9月5日在哈尔滨工业大学所做的讲座整理而成。

We assume that poor people all wanting to live a better life. But sometimes we don't know if the people want to move or not, and here is one of the famous occasions in China that all have left but one resident did not want to, so the new big building were built around the small house. When we clear things very fast, people are unsettled by the changes. When we change something and demolish places, it takes away a lot of memories. We expect that people from the countryside start adjusting quickly to new places and new opportunities. Although a lot of them become construction workers and get other jobs, the transition cannot be made overnight and it seems that there is still a wall between previous life and new life.

When we demolish all the poor areas to build new buildings fast that may offer a lot of new opportunities and homes for new arrivers, we concentrated on economics only. But research tells us that Social and Culture infrastructures are as important as the economic one that the people who are displaced from the poor areas often do not find happiness. So a balance must be found in the new neighborhoods between these three considerations if sustainable places are to be achieved where people can live in happiness.

Most western economists identify three stages of economic development: the first is production, making of things; the second stage is consumption of all these things until all are lifted out of poverty, and then the third stage when all have enough and can move towards a stage of civic enlightenment. But one thing the economists haven't recognized is that people never have enough and will always try to have more even when we don't need it. "I shop there I am" has become a way of life in an era of consumption .

And this consumption is producing a lot of waste so that we can consume more. These slides are trying to show how much we wasted by having more then we need.

Wealthy people are throwing 30% of food away although there are so many people who are starved. Some areas of my city, 4 % of population owns 70% of the wealth . Few people have a lot and many people have very little. People are building walls and gates to separate the wealthy from the poor and this is not a good way to build happy communities.

The other challenge is that we are using too much resources, about three times as much as the earth can sustained for our generation. The big problem is that we have produced too much CO_2, and this is changing our climate. We just have to keep on eyes on the disasters around the world to see how this is creating social problems all over the world. And the biggest problem is shortage of water which is getting worse than ever. Some of the people who hear about the bad things happening on earth think of it as happening far away. But the glob is connected and a disaster in one part of the world has an effect on the rest. So it is that it's time for us to change the way we are thinking. We can't do things like we have always done them because it's damaging our future.

And many people, including architects, are trying to design for less consumption and less waste.

Here is a new neighborhood design in Kroensberg, Germany. The architecture changes a lot, (as in the pictures) , new solar technology used on the surface of walls and roofs making a lot of energy for the neighbourhood , new ways to shade the building with plants and new material are making the building more beautiful. New electric trams to the city are faster and cleaner. Cycling in the city is always nice and good way to be healthy, although it's too cold in winter. But in China, it's a sustainable way to solve big problems in the city of medium distances. Copenhagen, Denmark, is producing a 20 years' plan to keep the car out of the city instead of bike. And it's a truly inspiring way to change the behavior of the modern life. And another way to save resources is to return to shopping in small, independent shops and markets rather them malls with international companies. Small shop shopping is always a pleasure and it saves energy and the environment. It is also a good way to retain money in the local economy.

Most architects don't think that we can respond to these big changes and we think that it's just somebody else's problems. But we do have responsibility because we have knowledge. We really need knowledge to gain real understanding of all these things. You have knowledge and you have the responsibility to lead the new solutions.

When you start working, there are two ways you can choose to practice:
One is to do more of the same. You build big buildings fast and not think of any of the problems it produces. You can go on with no worries, and no feelings of responsibility. All these pictures show the architecture that is produced by the architects who don't care. Although we are trained to do beautiful architecture, we are still not the main building makers in the city. Architects love taking prizes, and it's a really small part of professional results that does. Most architects design very poor buildings that damage the planet.

The other way is to be a part of the solutions. It will become big challenges, and you will find it is hard to find way out. But it shows in many different ways that one man can make an effort to change the world. It is the way to change the way people think. A lot of international architects are trying to build a city with Zero carbon omissions, and find a way that we can still continue having beautiful buildings but not taking too much resources. Your pizza hut is exactly the same as ours in London, and it's quite easy for us to change all of these things in some short distances.

In China, there are some really nice projects that are trying to push all of the design and the way of thinking to be more international. This one has been recently published in England, and we are really interested in the high density of a new way to achieve a building.

Here is one solution in New York. Their mayor has decided to achieve a better New York, they painted new direction marks, and make more public spaces for people to have a good time. If you want to do something for your city, there is no need to make a rule or any strict rule. It will be a start for people to change.

Well, so what does it mean? If you want to find a better way to build, you really have to work in a different way. It has to start from the education of the young architects. One profession can't change the world, but we can help to visualize alternatives. We have to think more of well being, and to think less about making money. These are not some normal shifts of values that will happen overnight - it could take a long time which we don't have. It's for you to get on with an amazing design and search to show the rest of the world how it should be done.

So the future is up to you: what will make you get out of bed in the future?

The Making of Eco-homes in England[1]
英国的生态住宅

Lucy Cartlidge
School of Architecture, University of Sheffield
谢菲尔德大学建筑学院

Definition

Guy and Farmer's assertion that the terms 'sustainable housing' and 'eco housing' in practical usage cover a multitude of meanings and confusions and an array of eco-logics of sustainable architecture which are often used interchangeably without a consensus as to what constitutes an eco-architecture (Guy and Farmer 2001).

The Code for Sustainable Homes

2006 (December) Launch of the Code for Sustainable Homes
2007 (April) The developer of any new home in England can voluntarily choose to be assessed against the Code
2008 (May) Mandatory to obtain a rating against the Code

Objectives

To understand the making of the meaning of 'eco-home' through the making of eco-homes.
To understand the making of eco-homes as relational, that they are not made as separate objects to us as human beings but that in making eco-homes there is also a re-making of ourselves.
To understand the re-making of the eco-home by residents into something that may be more or less eco depending on the residents' background or desires and thus re-making

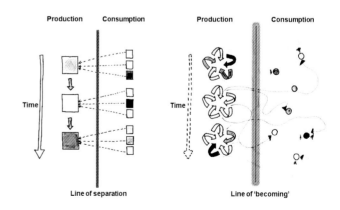

---

1 本文根据Lucy Cartlidge于2012年9月5日在哈尔滨工业大学所做的讲座整理而成。

Concrete / measurable

Theories of Relationality

Stuff
Building Regulations, Carbon trading, Renewable Heat Incentives, Feed in Tariffs

Dualism between nature and culture, limited discussion of values and meanings, focus on material culture and practices

Material Culture, Socio-Technical Studies (STS), Consumption, theories of 'practice'

Shove – normalising of behaviour such as showering and kitchen layouts

Kaika – domestication of nature in the household

Shove and Warde – inconspicuous consumption of water

Roger Silverstone – domestication, taming technology and transforming relationships

Doing
Sustainable Communities, Congestion Charge, Recycling incentives, DEFRA Sustainable Lifestyles

Dualism between nature and culture, theories seek to bridge this divide and focus on values and meanings but still acknowledge a differentiation

Barr – quantification of recycling and number of flights, categorisation of environmental behaviour

Szerszynski – technocratic campaigns of WWF and Greenpeace need situational meanings

Dobson – political identities and environmentalism

Clayton and Opotow – relationship to nature, environmental identity

Thomashow and Weigert – deep green, an eco-identity

Simmons – environmental ethics, instrumental or intrinsic

Being
Racial equality
Asylum and immigration

Non-dualism, theories seek to demonstrate interrelationships between nature and culture, focus on experience, values and meanings

Essentialism

Plumwood – modernity and enlightenment led to dualism between science and experience which resulted in over-use of resources and pollution

Pratt, Howarth and Brady – environmental crisis stems from inauthentic being

Sharr – Heidegger and architecture, theories of dwelling

Merleau-Ponty – 'being-in-the-world', interbeing

Ingold – dwelling perspective

Eco-Being

Abstract / ephemeral

Dualism ⟵⟶ Non-Dualism

Pair of earth sheltered houses
- Earth sheltered
- Geothermal
- Natural lighting
- High thermal mass
- Turf roof
- Eco-aesthetic

The Wintles, Shropshire
- Passive solar heating
- Natural ventilation
- Natural lighting
- High spec insulation
- Vernacular materials
- Embodied energy
- Energy efficient
- Growing vegetables
- Eco-cultural / eco-social

Allerton Bywater Millennium Community
- High specification insulation
- Modern methods of construction
- Energy efficient
- Eco-technic

the meaning of 'eco-home' in relation to the re-making of themselves.

Case Studies
- vary in scale
- all under construction
- all in England
- 'eco-housing'
- 520 dwellings at a Millennium Community in West Yorkshire
- 40 dwellings by an inexperienced developer in Shropshire
- 2 earth-sheltered buildings in West Yorkshire

Model
"The Millennium Communities Competition at Allerton Bywater will provide a model for sustainable development capable of wider application in other settlements where the creation of more sustainable lifestyles for new and existing residents is the key objective. The Millennium Community will take its inspiration from the quality, vibrancy and unity of village life and match this with innovation in building design and sustainable development as a model for 21st Century communities."
(English Partnerships, 1998, Millennium Communities Competition Allerton Bywater Stage Two Development Brief, p.5)

Confusion
However, there was strong opposition from the local community due to the high density of the build (Akilade 2000) which would see their village increase from 1580 households (English Partnerships 1998) to over 2100 households.
"The developer consortium [...], consisting of [...] and [...], was accused of not explaining the benefits of the scheme to the residents. Some in the community were not ready for the innovative features planned for the site, and as a result many of the modern features were toned down and the homes made more traditional in appearance" (Unknown 2003).

Self–build
"The Trust has decided to sell plots on the site for self-building. To some people, 'self build' means literally dangling in every last nail, whilst others it means commissioning an architect and builder to complete a house to order. To us, 'self build' means that every new dwelling on the site will have the participation of the householders, thus creating a living environment that accommodates the needs and aspirations of the inhabitants as opposed to a standard housing estate..."

Motivation
Question: What inspired you to want to create this eco-friendly house?
Eric: Eco-friendly house. What made me do it? I actually saw one, when we did another project a few years ago and I thought this was the way forward, and then when I looked into it more. And I kept getting fuel bills, I thought, that is the way forward, instead of building wind mills and building nuclear power stations, and all screaming out for energy, stop using the energy and if you can stop using it we shan't need them shall we?

Environmentalism
Eric: Yeah, I'm not an eco-warrior or 'owt daft like that. But I'm passionate that we are ruining planet, yeah, I really am, ahh.
Question: Do you think it is man-made?
Eric: I really do, I do. Then I said to myself, we've had ice ages. I were at [...] last week, [...] house, doing a job and you look over Sheffield, it's a fantastic view and it were like nine o'clock in the morning and I were looking at this job, this roof that we were going to do and I thought that's funny, what's that and there were a yellow band in sky all over Sheffield.

Exemplar
"unique within the borough and nationally unusual... as far as can be established, no earth dwelling occupies the site comparable with [this location]. If [this building] were to be developed then an earth dwelling is possibly the only solution

capable of responding to the harsh climate without impacting on the landscape. [This project] is therefore unique in that it is the only earth sheltered dwelling to occupy an elevated moorland site."

Pragmatic

"I don't think we'd have ever thought of having a grass roof… it just wouldn't have crossed our mind but because of how it's done now the field will always look like a field, and that is what we wanted it to look like. Just because we are going to build there we don't want to change what it is already there already and think oh, I'm going to make this into an urban garden."
(Viv)

Inspiration

"Ironically, whilst the impact of this particular dwelling will be very limited, quality architecture deserves to be seen, particularly so if it is to be held as an inspiration and example to others."
(Local Authority Planner, 2007, 12/04)

Conclusion

1. Policies and codes are limited by the way that they categorise the eco-home. They portray a limited vision of what an eco-home is. Firstly by considering it as an object which is quantifiable. Secondly by not taking into consideration other things which may form part of an eco-home: use of materials such as earth or straw, passive solar design, space for growing vegetables, craft-based production. Many of these things have been associated with eco-homes in the past but are difficult to quantify and measure.
2. Motivations for making eco-homes are not always noble. There may be pragmatic reasons why an eco-home is made that does not relate to environmental concerns. Planners and architects have a significant role in enabling and encouraging developers to do more.
3. The introduction of codes and ratings is positive as both tangible projects have been developed to higher environmental performance standards than they would have been. However the introduction of Codes and ratings has also changed the perception and awareness of eco-homes.

Architectural and Urban Acoustics: Approach Across Science, Engineering Social Science and Art[1]
建筑与环境声学：跨科学、工程、社会科学及艺术的探索

Jian Kang
School of Architecture, University of Sheffield
谢菲尔德大学建筑学院

研究团队 (Research Team)
Professor : 1
Visiting professor : 1
Postdoctoral RAs : 6
PhD. researchers : 15
MSc. students : 5
Visiting scholars : 8

研究网络 (Research Network)
Chair of EU COST Network on Soundscape
Chair of WUN Environmental Acoustics Network
Joint-chair of UK NoiseFutures Network

研究设备 (Research Facilities)
· Small anechoic chamber for scale modeling
· Large semi-anechoic chamber
· Rreverberation chamber
· Sound transmission suite
· Software for indoor and outdoor acoustic simulation1

1 声学：科学及工程 Acoustics: Science and Engineering

1.1 城市声学 Urban Acoustics
计算机模拟：辐射度模型 (Computer Simulation: Radiosity Model)
· Simulate sound propagation in various indoor and outdoor spaces with diffusely reflecting boundaries.
· The radiosity method was originally developed for the study of radiant heat transfer. While the application of the method is usually for steady state, such as in lighting simulation and computer graphics, an important feature of using the method in acoustics is that the time factor should be considered, which can increase the computation time significantly.
· Each boundary is divided into a number of patches, and the sound propagation in a space is simulated by the energy

---

[1] 本文根据康健教授于2012年9月3日在哈尔滨工业大学所做的讲座整理而成。

exchange between the patches.
・Also compared and combined with image source method, and ray/beam-tracing

城市街道的声学模拟 (Acoustic Simulation in Urban Streets)
Parameter study about effectiveness of architectural changes and urban design options–important being at the School of Architecture
Comparison between diffuse boundaries and geometrically reflecting boundaries

城市广场的声学模拟 (Acoustic Simulation in Urban Squares)
城市肌理的影响 (Effect of Urban Texture)
大尺度的噪声地图 (Large Scale Noise-mapping)
噪声地图的准确性和效率 (Accuracy and Efficiency of Noise Mapping)

城市声学可听化 (Urban Acoustic Animation)
・Various urban sound sources
・Dynamic characteristics of the sources
・Movements of sources and receivers
・Calculation speed reasonably fast

街道"峡谷"中的树木及植被：降低噪声的可持续手段 (Trees and Vegetation in Street Canyons: Sustainable Means for Noise Reduction)
More efficient due to multiple reflections, in terms of diffusion, transmission and absorption
Also considering landscape, air pollution, air flow, etc

London underground project

1.2 室内声学 Room Acoustics
长空间声学及其在地铁中的运用 (Acoustics of Long Spaces and the Application in Underground Stations)
・Acoustic theory for long spaces
・Computer models
・Acoustic scale modelling
・Speech intelligibility

地下购物街 (Underground Shopping Streets)
用餐空间的声学 (Acoustics in Dining Spaces)
A design merely based on the current guidelines for space use may lead to very poor conversation intelligibility –New design guidelines in terms of material, shape, seat arrangements, etc

CAAD 可听化 (Adding Instant Acoustic Response to CAAD-Animation)
古代表演空间的声环境 (Acoustic Environment of Ancient Outdoor Performance Spaces)
医院声学 (Acoustics in Hospitals)
NHS project

Acoustics modelling
Agent based model
Grounded theory based interview
Before-after study

Hong Kong new airport underground project

1.3 建筑声学 Building Acoustics

研究与示范项目：与"整座住宅低能耗"相匹配的低能耗通风系统 (Research & Demonstration Project: A low Energy 'Low-energy Whole-house' Ventilation System)

气密性和隔声 (Air-tightness and Sound Insulation)

绿色屋顶的隔声 (Sound Insulation of Green Roofs)

利用固体自由曲面加工技术的吸声装置 (From MPA to Strategically Designed Absorbers Using Solid Freeform Fabrication Techniques)

Girli 混凝土 (Girli Concrete)

2 声学：艺术及社会科学及心理学 Acoustics: Arts and Social Sciences and psychology

声景及声音舒适度 Soundscape and Acoustic Comfort

声景及城市开敞公共空间的声音舒适度 (Soundscape and Acoustic Comfort in Urban Open Public Spaces)

♦ Relation between human being and acoustic environment
♦ Based on over 10,000 interviews across Europe
♦ relationships between measured sound level and the subjective evaluation of sound level and acoustic comfort
♦ sound preferences
♦ main factors that characterise the soundscape

用语义细分法描述声景特性 (Characterising Soundscape Using the Semantic Differential Technique)

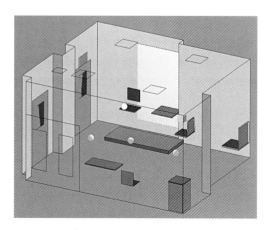

Simulation of hospital acoustics

在城市声学评估中神经网络的运用 (Application of Neural Network in Urban Acoustic Evaluation)

声学掩蔽 (Sustainable Acoustics for Comfort: Masking)

声景设计的艺术 (Art of Soundscape Design)

环境噪声屏障中的非声学因素 (Non-acoustic factors in the Design of Environmental Noise Barriers)

Public participation
Design process
Perception
Sustainability analysis

"非声学"空间的声音舒适 (Acoustic Comfort in 'Non-Acoustic' Spaces)

A series of MSc and BA dissertation
♦ shopping mall atrium spaces
♦ library reading rooms

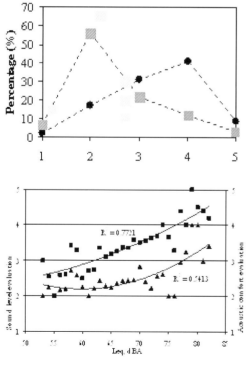

Evaluation of Soundscape

Comparison between the subjective evaluation of sound level (circular symbols) and acoustic comfort (squared symbols) in the Peace Gardens, Sheffield.
1, Very quiet (very comfortable); 2, quiet (comfortable); 3, neither quiet (comfortable) nor noisy (uncomfortable); 4, noisy (uncomfortable); 5, very noisy (very uncomfortable).

◆ football stadia
◆ swimming spaces
◆ churches
◆ dining spaces
◆ railway stations

开敞式办公室的声学 (Acoustics in Open Plan Offices)
室内空间中听力障碍学生的声舒适 (Acoustic Comfort and Sound Preferences of Hea ring Impaired Students in Indoor Spaces)
其他跨学科题目 Other Cross-disciplinary Topics
声学及可持续发展 (Acoustics and Sustainability)
气候变化以及建筑设计：(Climate Change and Building Design)
外部空间的宁静——声学和视觉因素的影响 (Tranquillity of External Spaces ——Influence of Acoustic and Visual Factors)
英国史前遗址声学及音乐元素 (Acoustic and Musical Elements of

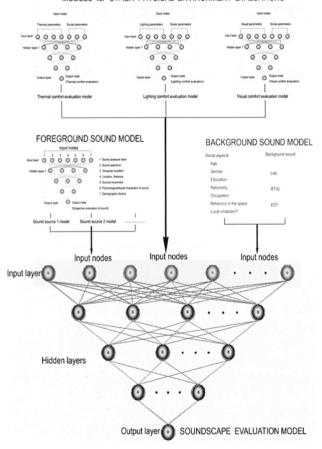

Neural network forecast of soundscape

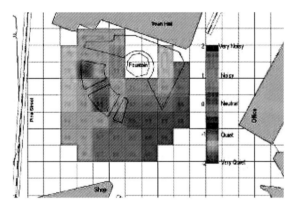

Age group 13–18

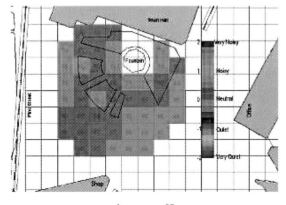

Age group >65

Soundscape mapping

Prehistoric Archaeological Sites in Britain)
河的声景 (River Soundscape)
Arranging recreation and resting spaces close to a river
— Build water features e.g. artificial extension from a natural river that can be switched on at days and off in nights
— Planning living rooms facing the river side while bedrooms– the opposite site

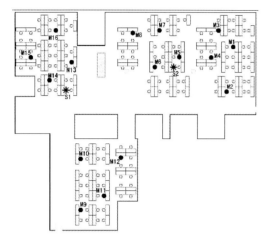

Royal Society funded project — collaborated with China
Introducing psychoacoustics into room acoustics

Acoustic effects of sustainable measures

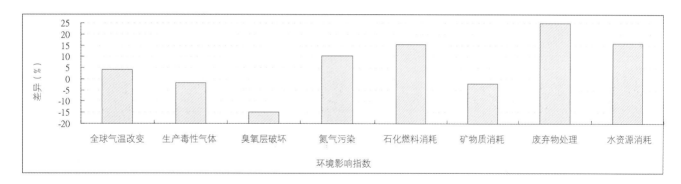

Sustainability analysis of acoustic materials

联合设计

Joint Design

操作 Operation

01

Future Scenarios

Constructing for the Unforeseen in Harbin

Foreign Teacher :	Dr. Renata Tyszczuk	
Chinese Teacher :	Associate Prof. Luo Peng	罗 鹏 副教授
	Lecturer Meng Qi	孟 琦 讲 师

Students :	Yue Chao	岳 超
	Qiu Lin	邱 麟
	Ma Xinran	马欣然
	Wang Rufei	王如菲
	Ge Jiale	葛佳乐
	Gao Bo	高 博
	Mao Yue	毛 悦
	Ding Fengming	丁凤鸣
	Zhao Hongchuan	赵洪川
	Cui Xiao	崔 潇
	Fan Lu	范 璐
	Zhang Xiran	张晞然
	Wang Hui	王 辉

2012 September

Academic Activity : International Collaborative Teaching and Acacemic Month
Institutions : Harbin Institute of Technology and The University of Sheffield

未来场景
——哈尔滨的意外建构

任务背景

城市相当于一个综合拼图，无论是人类生产生活的变化，还是自然地理气候的变迁，都关乎一个城市的未来发展。我们生活的城市将会以怎样的姿态展现于世？一个影响因子的变化，对城市的辐射面有多大？这些都是值得人们去思考与探索的。根据现有的条件进行多方面预测，对未来做设计，让决策者和设计者对城市的动态变化有一个总体的把握。对于积极因素，将其发扬光大；对于消极因素，要防患于未然；进而使我们的城市发展更贴近人们的生活与需求。

City is equal to comprehensive puzzle, no matter how human' activities and physical geography climate change, they all influnce the future development of a city. What is the future appearance of our city to the world? How large the radiating surface is if an element is altered? All of those things are deserved to be considered and investigated. According to the existing conditions, make various predictions and design for the future to enable decision makers and designers to have an overall grasp of urban dynamic changes. On this basis, positive factors are carried forward and negtive ones are prior restrained, which make our city developed closer to people'daily life and demands preferably.

作为未来的建筑设计师，需要时刻了解周边事物的动态变化，并相应地作出积极回应，给予正确的引导。这不仅要求建筑师对变化有敏锐的洞察力和设计解决能力，而且需要时刻保持知识的更新，不断拓宽视野并提升与他人的团队协作能力，把握城市发展趋势，为共同的生活环境质量的提高做出贡献。

As future architecs, we should be familiar with dynamic changes of surrounding things and give a positive response, which requires keen insight and resolution ability, also updated knowledge, broadened horizon,and cooperation ability with others as well. furthermore, grasp the trend of urban development whenever necessary to make a contribution to our promotion of living environment quality collectively.

课程目的

设计未来：
建筑设计其实就是对未来的设想，哈尔滨的未来是什么样子的？本次设计将在全球环境变化的大前提下来研究哈尔滨的城市未来形象。设计期间，时刻处于动态中的哈尔滨城市环境与全球的经济、生态、文化和科技等的发展是设计考虑的重中之重。
基于此，本设计训练旨在为哈尔滨市某区域未来20年内的发展变化提出一个设计策略。

Design for future:
Some times, architecture design means future imagination, so how about the future in Harbin? This mission is to investigate the future image about Harbin in the context of global environmental change. During this curriculum, the dynamic environment of Harbin, economy, ecology, culture, and technology development of global are the most important factors to be considered.
Based on this situation, this design task is aim to promote a strategy for some areas of Harbin about the next two decades.

设计关键词：食物，贸易，垃圾，气候，地形，所有权，政治，生态，文化，技术

Design category:
FOOD/ TRADE/ WASTE/ CLIMATE/ TOPOGRAPHY/ OWNERSHIP/ POLITICS/ ECOLOGY/ CULTURE/ TECHNOLOGY

课程特色

1. 训练学生以一个关键词切入设计主题，并以发散思维引申拓展出多种可能性，对自己的研究方向形成一类思维库。发扬团队合作精神，每个人针对其他所有人的设计进行头脑风暴，将自己的思维库移植到别人的设计中，集思广益，使不同人的思想融合在同一个设计中，在此基础上取人之长补己之短，完成最终的设想方案。每个设想方案体描述着城市某一方面的发展，而不同方向设计策略的集合则形成了阐释城市未来发展的百科全书。

1. Training students to start design by a key word through divergent thinking to expand various possibilities so form a special brain trust. With team spirits, brain storm is being used in everyone's project, transplanting one's own brain trust to others' design, making an integrated thoughts, then make the best to finish their final projetcs effectively. Every strategy is an illustration about one aspect of the city development, then when they are together organically, miraculously an encyclopaedia of Harbin is accomplished.

2. 训练学生学会运用蒙太奇手法进行建筑与城市学科方面的设计与表达，建筑设计是对不同设计思维的剪辑拼贴，未来城市则是不同时空下的建筑的有机合成。

2. Make them get familiar with the mathod that use montage to express design idea of architecture and urban design subjects. Architectural design most like a cutting and collage of different thinking, and the future city is an organic synthesis of architectures in different time and space.

左图为全体同学和老师们的合影

Left is photo of all the students and teachers

课程日历

DAY	TITLE	KEY TASKS	OUTPUTS	YOU WILL NEED
	Preparations before Class	TASK **1 A1** — Prepare a plan drawing of a place in Harbin in any scale (simple black and white line drawing). This can be a building, street or larger area, randomly selected from a map of Harbin	Summary of your site' advantages and disadvantages	
		3 A4 — sheets of the entries for their assigned category in the **Encyclopaedia of Harbin**	You will bring these 3 papers on Monday to share your opinion to the other	
Monday	Make Jigsaw Map and Draw the Future of Harbin	TASK 1 Give self-introduction to others and make a brief interpretation of your chosen site.	Getting to know each other and where your site it is.	• Chalks • Some colorful pens or pencils • Camera • Some blank papers
		TASK 2 Use chalks to draw a map of Harbin on the floor, including the scope of all sites. And begin to imagine what Harbin will be like in the future and draw them out directly.	locate each site plan on the map correctly and make some head races from Songhua River to each site to show our future city.	
		TASK 3 Describe your category and give some special opinions about it to all.	Make others and teachers fet familiar with what it means to Harbin.	
		TASK 4 Take turns deliverring each site plan to drow your own thoughts on it.	Brain storm to collect all ideas together on each design, making work more efficient.	

DAY	TITLE	KEY TASKS	OUTPUTS	YOU WILL NEED
Tuesday	Select Useful Thoughts and Rapid Design	TASK 1 Neaten one's own work seriously and begin to design for site according to the assigned category.	Give out a strategy to solve the problems related to the site in any media.	• Modeling equipments and drawing materials • Reference pictures
Wednesday	Give Guidance to Design	TASK 1 Make a discussion with each other about your own design and show your project to teachers	Receive feedbacks from your companions and teachers	• Modeling equipments and drawing materials • Reference pictures
Thursday	Deepen Project Design	TASK 1 According to the preliminary idea to interpret final project in montage	Final blueprint in A2 paper	• Modeling equipments and drawing materials • Reference pictures
Friday	Final Presentation	TASK 1 Individual presentation with blueprint TASK 2 Teachers give comments and course summaries synthetically	Course is successfully completed.	• Final project • Camera

作业一

岳 超

参与感想：我们用了将近2个小时的时间将每个人的改造基地拼到了一起，汇集成了未来哈尔滨城市地图意向，在这个拼图中，我们用手中的笔对其进行宏观的规划与构想，很有意思，学到了很多课堂上没有的知识。

Reflections : It takes us nearly 2 hours to match every one' site together, forming a image reflecting the future of Harbin,then we use pen to draw our thoughs on it directly. Such an interesting task!

设计说明

所选主题为食物，即我们吃什么，种类是什么，来自哪里？农业、畜牧业、传统的、可持续的……
基地选址为道外区的一个广场，在2026年这里将举办盛大的美食节，美食节举办后这里会发生什么，变得如何？让我们拭目以待。

FOOD that means what do we eat from agriculture, stockfarming, traditions, sustainability……
This picture is my final result of this work, and that is a square in Daowai District and here is the image about the Food Festival when it held in this place in 2026. So what will be happened if the Food Festival be held here?

切入点1：农业与水产业
1. 水稻种植基地——香坊区东方红农场
2. 蔬菜种植基地——道里区
3. 水果种植基地——南岗区
4. 禽畜饲养基地——太平区青年农场
5. 捕鱼基地——松花江

1. The grain planting base—Xiangfang District, named Dongfanghong Farm
2. The vegetable planting base—Daoli District
3. The fruit planting base—Nangang District
4. The livestock base—Taiping District, named Youth Farm
5. The fishery base—Songhua River

切入点2：佳肴与特产
1. 哈尔滨红肠
2. 老鼎丰：哈尔滨著名的月饼
3. 大列巴：原产地俄罗斯，外表硬脆，内里松软
4. 哈尔滨啤酒：哈尔滨人都会喝
5. 马迭尔冰棍：有100多年的历史

1. Harbin sausage
2. The cake of Laodingfeng
3. The leba bread : from Russian, the external is hard while inside is soft
4. Harbin beer
5. The ice cream of Madieer : enjoy a history of over 100 years

切入点3：文化
1. 中华巴洛克：这是一个巴洛克历史文化保护区，覆盖道外头道街至20道街范围。
2. 调研：景阳街两侧的老建筑拥有大量复杂的线脚、山花和装饰符号，也许不适合现代生活，但有历史价值。
3. 哈尔滨有大量稀有的巴洛克老建筑。

1. The baroque in China : ranging from Daowai 1st street to 20th street.
2. Investigation : massive and complicated skintle, pediment and some decorative pattern are in the facade of those old buildings beside Jingyang Street.
3. Rare historical buildings of Baroque style in Harbin.

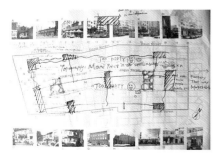
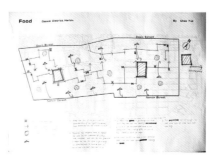

制造一些距离现有著名的餐馆很近的路径穿过这个大的街区，便于人们发现它们，同时提供多变的空间。

Some paths are built to go through into this big block to provide various space for residents.

作业二

邱 麟

主题阐释：贸易，是自愿的货品或服务交换 。随着社会的发展和科技的进步，贸易目标从实际延伸到虚拟，市场从有形拓展到无形，不断丰富和进步。贸易也是商业，大到国家与国家，小到个人与个人之间，都会有贸易出现。为方便贸易活动的开展而出现的设施也变得多样化起来，这是由贸易带动的对城市未来发展的可视变化的结果。

Reflections : TRADE, which means goods and service voluntary exchanging. With the development of the society and technology, Trade Objectives expands from reality to virtual world, as well as the market. Trade is equal to business, trade will exist no matter for nations or persons. Facilities serving for trading also changed variously, which contributes to the development of our cities.

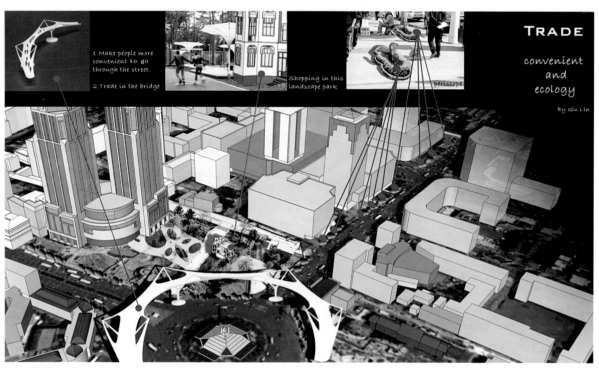

设计说明

在红博广场建造过街天桥，缓解交通压力，为行人带来便利，且在不久的将来定会形成一条很有特点的空中商业街。让老建筑重现天日，并在其周围形成一片树林，既能改变周边的小气候，又能让购物的人们享受绿色空间。为地下商业街提供一种设施——潜望镜，方便地下购物的人们随时定位。

Build overpass for Hongbo plaza is aimed at easing the traffic pressure and bring convenience, and it will form a special commercial street in the air in the near future. Then represent the old building and form a forest to change microclimate and offer a green space either. Last set some telescopes for subterranean commercial area to help them easy locating.

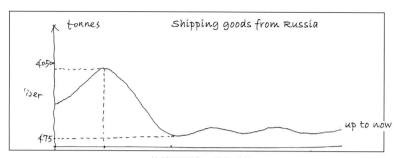
从俄罗斯进口货物曲线图
Shipping Goods from Russia

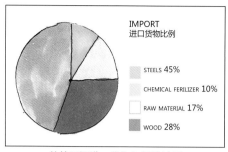
从俄罗斯进口货物比例饼状图
Total Goods From Russia Since Sailing Open

哈尔滨经济贸易洽谈会

秉承"突出俄罗斯、面向东北亚、辐射全世界、服务全中国"的办会宗旨，哈洽会在俄罗斯及东欧、东北亚国家和地区具有较大的影响力和知名度，每年6月15日至17日在哈尔滨举行。

HTF
(Harbin Economic and Trade Fair)
Adhering to the aim of 'Facing to Russia, Northeast Asia, and serving China as well as the whole world', HTF enjoys the enormous influence and popularity among those countries.
The annual session of HTF is held in Harbin National Convention Center from 15th to 19th of June.

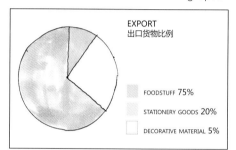

交易场所形式
Form of Trading Place

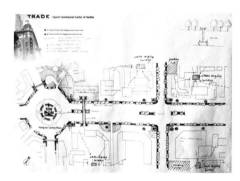

| 精品店 | 个体书报亭 | 商业街 |
| Boutique | Kiosk | Commercial Street |

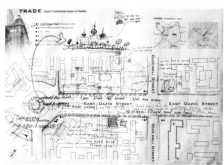

作业三

马欣然

垃圾——该地段地处大学旁边,交通混乱,但是人气很高,仅有的小吃街已逐渐无法满足人们的需求。该设计希望能通过步行街改造,对原有的区域功能进行丰富和空间重组。把这一区域打造成2021年富有活力的步行娱乐圈。

WASTE—The site is located besides a universe, enjoying a high-popularity even though its aggressive traffic. However, the only few of snack streets cannot satisfy people's needs, so this project wants to enrich and recombinate the original founction areas to form a most active pedestrian street in 2021.

    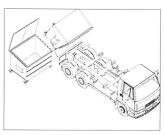

| 居民垃圾 | 居民楼竖向垃圾道系 | 垃圾收集 | 垃圾倾倒 | 垃圾运输 |
| Residential Waste | Refuse Channel in Building | Waste Collection | Waste Disposal | Waste Transportation |

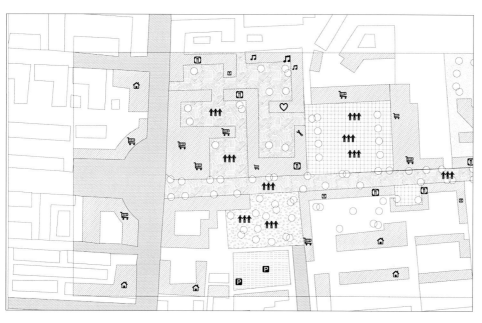

本设计通过区域的步行系统设计，试图解决原有街区的问题，并通过对周边功能的改造和活化，使其更符合周边学生和居民的娱乐需求。

This project wants to solve the problems left in the blocks, and transform founctions around to meet the entertainments needs of students' and residents'.

虽然课程时间不长，但是我感受到了新颖的教学模式和作为建筑师责任感的重要性，认识到了细节决定成败。

Though the time of course is not enough, I could catch the point of innovative teaching mode and how the responsibility important for an architect, and is realising success depends on details to some extent.

作业四

王如菲

气候——20年之后可以将松花江从地下引入城市，不仅在夏天可以降温，冬天可以举行冰雪文化活动，带来宜人亲水的城市环境，同时可以利用水景建立水循环中心，从能源的角度进行水的循环可持续利用。

CLIMATE—We could make water diversion from Songhua River to our city 20 years later to feel more cooling in the summer and hold more activities theme by snow and ice. Then the new area will bring pleasant hydrophilic environment for people and establish water recycling center. Finally we can realise the sustainable development of water cycle with an energy principal.

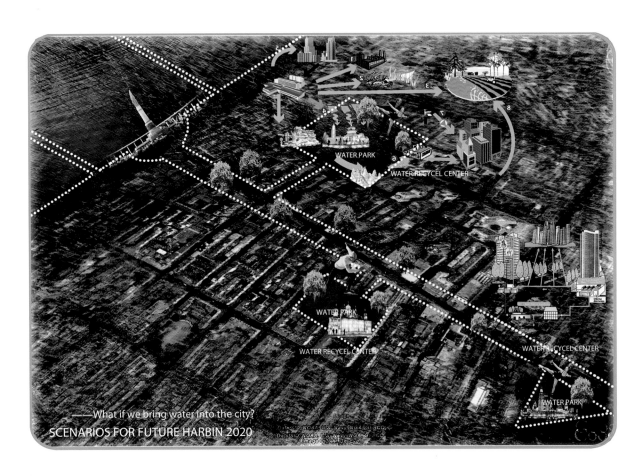

—— What if we bring water into the city?
SCENARIOS FOR FUTURE HARBIN 2020

现有绿化带分析	现有公共区域分析	现有广场分析	现有垃圾桶分析	现有水源分析
Green Belts Analysis	Public Area Analysis	Plaza Area Analysis	Garbage Can Analysis	Water Resource Analysis

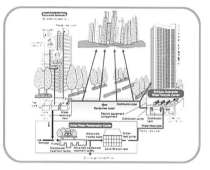 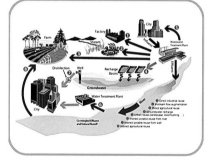 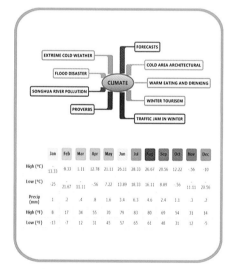

地下水循环中心补给建筑给水示意图
Groundwater circulation for building

地下水循环中心补给与城市、农田示意图
Groundwater circulation for city and farmland

地下水循环系统与城市、农田、建筑给水示意图
Groundwater circulation for city、farmland and architecture supply

未来哈尔滨设想图
Future Scenarios for Harbin

作业五

高 博、葛家乐

参与感想：设计是团队的成果，不只两个人，是更多的人在一起的奇思妙想。草图是设计的源泉，涂鸦和拼贴给了设计者灵感和梦想。

Reflections : Design is the result of the team, not only two people, is more and more people together of the idea. Sketch is the source of design, graffiti and collage inspired designers and dreams.

设计说明
所选主题为地质情况，松花江的流过造成了哈尔滨特有的地质环境和自然景观，流动的江水也在不断地改变着城市的地质状况。流经哈尔滨的松花江上有着大大小小若干个江心岛，在江水的冲刷下，10年、20年、50年、100年后会有怎样的变化？我们需要认真思考。

GEOLOGY—The Songhua River is constantly changing the Harbin geology status⋯
What kind of changes of the islands by the erosion of Songhua River in the future? That's what we need to think now.

切入点1：生态环境
1. 栖息在松花江周边的动物
2. 松花江沿岸的植物
3. 水生动植物
4. 城市对水域的需求
5. 市民所需要的生态环境

1. The animals along the Songhua River
2. The plants along the Songhua River
3. Aquatic organisms
4. The urban demand from the river
5. The need for public environment

切入点2：地质和生态的变化
1. 松花江水位不断上涨
2. 江心岛面积越来越小
3. 珍贵物种正在逐渐灭绝
4. 现存环境较为自然生态
5. 城市的绿肺

1. The rising of Songhua River's waterlevel
2. The Islands in the river are getting smaller
3. The extinction of valuable species
4. Ecology of the natural environment
5. Urban green lung

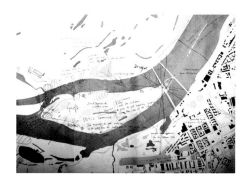

切入点3：我们的需要
1. 我们需要有桥
2. 城市居民需要一个生态公园
3. 我们要去适应江心岛的变化
4. 开发的同时要保护珍贵的物种
5. 开发建造不仅仅是一代人的事情

1. We need the bridges
2. Urban residents need an ecological park
3. We're going to adapt to the changes of the island in the river
4. Development while protecting valuable species
5. Development of the construction is not just for one generation

作业六

毛 悦

参与感想：我的分类是所有权。调研中，我对"所有权"从6个不同的角度分别进行分析，这些分析为我在第一堂课上快速分析其他各组的基地提供了很大帮助。大家针对不同角度的基地分析，帮助我快速地提炼出了该地区的3种发展的可能性，让我很好地完成了设计任务。

Reflections: Ownership, My topic is ownership, and I investigate and analyse it in 6 different perspectives, which do a great help for me to analyse others' sites quickly in the first class. With the different thoughs about my own site from others, I abstract 3 possible directions for my project. And so complete this mission perfectly.

设计说明

在最后的成果中，我利用拼贴的手法来表现我的设想。在地下空间，保留原有的商业街，将地铁站、商业街连为一体；另外，在地下设置小型的垃圾预处理设施和发电设施，以期分担该地区巨大的能源消耗和垃圾产生。

I use collage technique to achieve my opinion in finall work. Keep the original commercial street and make a link between subway station and business street. Besides set small garbage pretreatment facility and power generating equipment, to ease the burden of energy expenditure and the waste generation.

哈尔滨冬季气候寒冷，拥有大量的冰上运动，所以每年哈尔滨都会举办国际著名的冰雪节，自1985年起，迄今已成功举办了27届。夏季的哈尔滨会迎来盛大的啤酒节；同时，在这个季节，哈尔滨每两年举办一次大型音乐会，吸引了大量的艺术家前来。

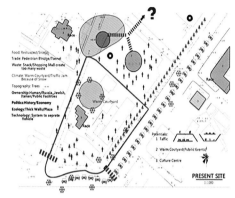

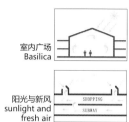

室内广场
Basilica

1. 创造一个室内广场，作为市民休闲、举办公共活动的场所。

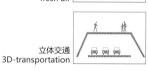

阳光与新风
sunlight and fresh air

2. 作为服务中心，为该地区提供必要的支持，如垃圾预处理、分担部分电力供应等。

立体交通
3D-transportation

3. 通过利用地铁建设的契机开发地下空间，建造过街天桥等方法创立立体的交通体系，以解决人车分流、堵车等交通问题。

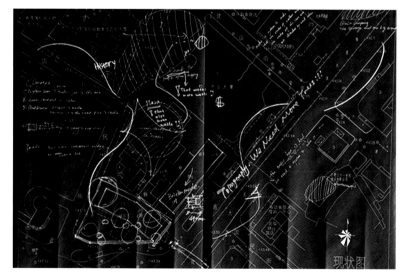

所选的地块位置特殊，周边商业设施众多。交通拥堵现象严重，需要大量能源和垃圾处理、污水排放、交通站点等公共设施的支持。如果在这一区域建立一个综合服务中心，为周边商业设施提供支持，城市会变成什么样？

作业七

崔 潇

主题简介：生态——漫游者计划
当前的哈尔滨城区绿化面积偏少，城市中缺乏供人们漫步休憩的公共空间。构想中未来的哈尔滨应是花园回归、绿色回归。设想以地段中的建筑学院为契机，构建一条漫步栈道；不仅联系街区，也是城市绿色花园蔓延的开始。

Reflections : ECOLOGY—ROVERS PLANNING
Nowadays, there is a obvious decay in green space and a lack of public space for wandering in Harbin. I imagine that Harbin should be a garden city in the future, as a result, I take our Architecture College as an opportunity to set a wondering catwall, which not only be a link between blocks, but a begining of the green garden spread.

设计说明

以"城市漫步"为出发点，将哈尔滨工业大学建筑学院作为所选地块的活力中心，从"绿色回归"出发，将三个相互独立的block整合为一体，使哈尔滨工业大学建筑学院作为中心点向周边发散出绿色的正能量。这种设计理念在未来可以持续影响到整个城市，使哈尔滨真正成为绿色花园城市，成为城市步行漫游者的天堂。

As a starting point of "cityeasy", chose School of Architecture, HIT as an active center, integrate 3 independent blocks together, making it as a green positive energy sending point. This design concept will affect the whole city in the future, transforming Harbin into a real garden city and a heaven of wandering.

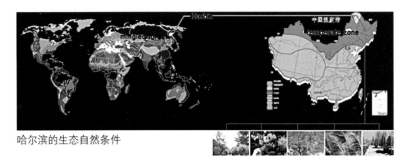

哈尔滨的生态自然条件

哈尔滨，东经125°42′~130°10′，北纬44°04′~46°40′，中温带大陆性季风气候，四季分明。

Harbin is in east longitude 125°42′–130°10′ and northern latitude 44°04′–46°40′, as well as in the temperate zone continental monsoon climate, having four distinctive seasons.

哈尔滨动物园饲养了很多物种，这些物种丰富了哈尔滨生态的多样性。

The zoos in Harbin raises many species, which actually increases her richness.

哈尔滨的植物大部分都是喜阳的植物，它们耐寒、耐旱。它们适应冬季的漫长、寒冷，夏季的短暂、凉爽。这种特殊的气候也让哈尔滨形成了特有的生态结构及建筑建构形态。

The plants in Harbin is mostly sun plants, cold resistance, and drought resisting species, all of which are adapt to the climate of a long cold winter and a short cool summer. this special climate allows Harbin to form a particular ecological structure and architectural style.

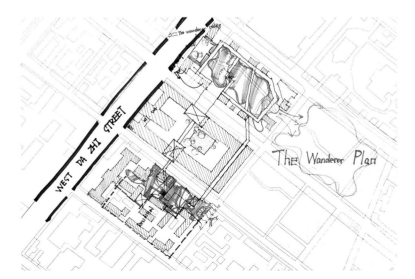

所选地段为哈工大建筑学院周围的三个街区。大家不约而同地将学院作为考虑的重点和地段的中心。给了我很大的启发的是：三个街区应该互相有联系的想法以及"绿色花园的回归"这一个主题。因此，漫游者计划的设想应运而生。

This site includes 3 blocks near my college, and my mates all take college as a key point and a center for design, giving me significant inspiration as follows.
Three blocks should have some relations and a vivid theme, so rovers is coming soon.

02

Collage Memory

— Fast Experssion

Foreign Teacher : Satwinder Samra
Chinese Teacher : Shao Yu   邵　郁
                  Dong Yu   董　宇

Students :   2010 级   24 名本科生

2012 MAY

Academic Activity : Oversea Joint Design Studio
Institutions : Harbin Institute of Technology and The University of Sheffield

拼贴记忆——快速表达

任务背景

刚刚步入建筑学专业的学生,在初学设计时所面临的最大挑战之一,便是寻找设计的原点。一个设计由何而起,是一个因人而异的过程。身边的一切事物、偶然发生的事件以及自己的生活经验,都可能成为一个设计的原点。如何开始一个设计,并从中整理出设计的逻辑,是此时此刻的一大难题。此外,在获得了灵感之后,迅速地、可视化地表达自己的想法,则是方案发展阶段的必要能力。学习建筑设计实际上是在学习思维的方法。这种方法论不是对思维的限制,恰当的思考方法恰恰能够为设计提供更多可能性。

When we first step into School of Architecture, one of the biggest challange is about finding the origin of a project. The question of how to start a project is quite different people by people. Everything around us would be the start point, just as some coincidents or even personal experience. It's quite hard sometimes to start a project and to work out a reasonable logic. On the other hand, after we get inspired, a quick illustration to show the idea is quite important when we develop the project. The process of the architecture design studying is more like the thinking method. The methodology is not a limit to think. Furthermore, a proper method of thinking can provide more possibilties to a project.

具体而言,"拼贴"这种表达方法在谢菲尔德大学建筑学院应用非常广泛。"拼贴"实际上是一种用已有的图案,通过重新组合,构建自己希望呈现的意向的表达手段。这种手段不仅可以用于前期方案的构思,方案和空间推敲,甚至在最终的成果表达阶段也很实用。由于运用已经存在的意向和图案,重新组合而成的图像具有较为真实的材质与质感,很容易表现场所的气氛。同时由于可以运用现有素材进行组合,大大省去了重新绘制的时间,节约了表达时间。

To make it more concrete, a kind of illustration called "Collage" is widely used in School of Architecture Sheffield. "Collage" is a mean of drawing, which we make a image by rearranging some exsisting pattens, showing the opinion we want to present. It is not only useful for the primary conception to show what we want when design. also helpful to the processing of thinking, even the final presentaion. Because of the using of existing pattens and photographys will be used, it is the atomsphere with the real textile well expressed. And in this way, the time to redrawing could be saved.

特别在方案推敲阶段,及时地将自己的想法进行可视化的表达,实际上可以加速思维过程。方案的推敲是一个手脑互动的过程。在画出一个新的建筑场景后,大脑会以此为基础,发现现有秩序中新的机会。所以,更快速地表达想法,意味着更快速地互动与思考。所以一种高效的表达手段是必需的。拼贴恰恰是这样一种快速、直观的表达方法,其熟练运用对于建筑学的思考是很有益处的。

Especially when we work on the primary conception, make visualized drawings in a quick way could contribute thinking process. Our hand work together with our brain when we desgin. When we work out a new scene, our brain will start to learn from it. In this way, a new thought or a diffenrent chance could be found. So a quick drawing skill is powerful. To use it expertly and intuitively in design conception will be helpful.

课程内容

本课程的主要目的是通过童年记忆的回顾，发现"非建筑"时期对于建筑和空间最原始的体验，以回忆与经验为基础思考建筑。此外，通过拼贴场景的训练，了解并掌握拼贴这一表达方法，并将其应用于之后的建筑思考中。

The main propose of the studio is to find ovringinal experiences of spaces when in the "non-architecture", by looking back to childhood memories. Moreover, students may learn the technic of collage by this training, which may be applied to future design process.

课程分为两部分内容。
第一部分内容：学生通过拼贴的手段，展现童年记忆中的一个场景。拼贴的素材为一本杂志中的图片。在拼贴的过程中需要特别注意视平线位置与透视关系，只有在视平线与透视正确的前提下，要素才能正确地形成场景，否则就不能构成一个空间。通过拼贴，即使是不会画画或者画画能力较差的设计者，也能通过重新组织现有的各个要素，轻松地构建出一个清晰的情景，便于交流。

The studio has two parts. The first, students use collage to present a childhood scene, by rearranging photoes or pattens on a magazine. The horizon and perspective is especially important in this work. The scene will be made sensiable only with correct horizon and perspective. By applying the technic of collage, those who have poor drawing can quickly construct a sensible scene to communicate.

第二部分内容：学生通过拼贴构建一个建筑场景。正如第一部分一样，所有拼贴的素材来自于杂志。首先将杂志上与建筑有关的要素扫描并打印，之后对要素进行重新组织，在一张白纸上拼贴出一个全新的空间。所用的要素可以是空间界面，如砖墙、拱顶、柱廊等，也可以是门窗等具体的构件。由于所构成的是建筑空间，所以正确的透视尤其重要。在创建了基本空间界面后，还可以通过拼贴人物提示建筑空间的尺度，并且用彩铅加上光影关系。通过这种拼贴的方式，学生可以很快地构造出所想到的建筑场景。这种拼贴场景，由于带有真实的材质，所以从中可以得出更加确切的空间体验，并易于发现新的关系。

The second part is about making a achitectural scene by collage. Just like the first part, students use pattens from magazine and rearrange them. The pattens could be architectural elements, like brick wall, arch dom, or even doors and windows. With correct horizon and perspective, students can attach some real people to show the scale, and use color pencil for light and shade. With this technics, students could present the architecture scene they imagined quickly, and this achitecture scene with real textiles can provide a more vivid experiecnces.

课程所展示的是一种快速表达的方式。通过拼贴，表达速度加快并且所呈现的可视化场景更加直观，展示着一种更易于感知的建筑空间氛围。对于方案过程，甚至最终表达，都是一种很有价值的尝试。

The Studio shows a quick way to present a idea. We can present a visible scene faster and more directly by collaging, and show a more sensible space atmosphere. It is a good technic both for our design work and do final presentation.

授课过程

儿时记忆（训练一）

- 个人的
- 经历和体验的反映
- 空间的和非空间的
  有关事件、地点、时间、空间

幻灯演示（讲授）

- 以往的+现存的建筑观念
  建筑设计需要"想象的能力 + 表达潜在未来的能力"
- 主题拼贴画
- 空间特殊性
- 图解性
- 或大或小的制作
- 多种媒介
- 利用图片素材 + 制作透视

制作（训练二）

- 用其他的透视图素材拼贴出你自己的空间设计
- 考虑光（影）
- 留意并反思拼贴制作时产生的意外效果
- 利用人+活动完善空间表达

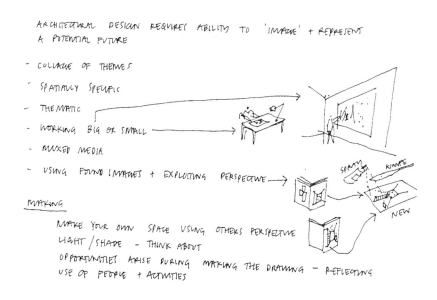

补充说明

- 学生可以利用绘画在拼贴上补全完善，也可以不画而直接表达
- 只可采用现实既有、可实现的元素
- 拼贴要具有三维空间的效果
- 互动检视——学生、导师、客户、使用者
- 制作可快可慢
- 展开并深化关于光、结构、材料、运动、居住的想法

徐筱铎
Xu Xiaoduo

李偲淼
Li Simiao

苏航
Su Hang

张天硕
Zhang Tianshuo

邵菁菁
Shao Jingjing

金艾伦
Jin Ailun

马忠
Ma Zhong

宋恩丰
Song Enfeng

刘凌灵
Liu Lingling

刘伊宁
Liu Yining

李偲淼
Li Simiao

张天硕
Zhang Tianshuo

邵菁菁
Shao Jingjing

游泽浩
You Zehao

孙惠萱
Sun Huixuan

马忠
Ma Zhong

03

Site Investigation

— Exploring London's South Bank

Chinese Teachers :	Associate Prof. Wu Jianmei	吴健梅　副教授
	Associate Prof. Luo Peng	罗　鹏　副教授
Foreign Teachers :	Dan Dary	

Students :	Wang Jinghui	王静辉
	Jian Rui	菅　睿
	Feng Zhe	冯　喆
	Tan Jiajun	谭佳俊
	Chang Xiangzhou	常湘珩
	Sun Yuxuan	孙宇璇
	Zhang Tianshuo	张天硕
	Wang Qiaochu	王翘楚
	Zuo Yu	左　煜
	Shao Jingjing	邵菁菁
	Xu Xiaoduo	徐筱铎
	Zhong Lanting	钟兰婷

Academic Activity : Oversea Joint Design Studio
Institutions : Harbin Institute of Technology and The University of Sheffield

2012 April

区域调研
——探索伦敦南岸

任务背景

在伦敦,相较于北岸,泰晤士河南岸地区发展得更为缓慢。由于南岸处于伦敦市的管辖之外,早在中世纪,这个地区便与那些在伦敦城内不被允许的世俗娱乐紧密相连。这些娱乐活动包括逗熊游戏、嫖娼以及戏剧。此外,著名的莎士比亚环球剧场就坐落在南岸地区。到了18世纪,随着沿河步行系统和舒适的公园的建设,南岸地区变得更加优雅宜人。然而很快,重工业的增长便取而代之。很多建筑架设于河上,阻隔了公众与亲水空间的接触。

The South Bank of the Thames River in London developed much more slowly than the north of the river. Due to its location outside the jurisdiction of London city, during the Middle Ages the area became associated with informal entertainment, not approved of within the city, which included bear-baiting, prostitution and theatre, Shakespeare's famous Globe Theatre was located here. During the eighteenth century the South Bank became more gentrified and respectable, with the development of riverside walking space and pleasure gardens. However, this was soon superseded by the growth of heavy industry, which took over the area, often building straight onto the river, cutting off the waterbank activity by public.

南岸地区在1951年的英国节时进行了完全的转型。此次转型旨在将伦敦乃至全英国从"二战"的破坏中振兴起来,例如皇家节日音乐厅这样的为英国节而开发的场地,将南岸地区重新定义为娱乐区。在这里,河岸区域重新对公众敞开。在之后的数年里,皇家国家剧院、海沃德美术馆,以及其他艺术场馆相继出现,并通过一系列的河岸步行系统和公共空间相连。在2000年,泰特现代艺术馆由原河畔电站改建并落成,这很大程度扩展了南岸艺术场馆的网络。

The South Bank was completely transformed in 1951 Festival of Britain, which aimed to reinvigorate both London and the whole country following the ravages of the Second World War. The development of the Festival site, which included the Royal Festival Hall, redefined the area as a place of entertainment, and this region opened up to the public once more. In the years that the Royal National Theatre, Hayward Gallery and other arts venues had developed, connected by a series of riverside walkways and public spaces. The Tate modern further extended the South Bank's network of arts venues.

任务目标

任务将会提供给我们一个机会去探索伦敦的一个地区,这里文化底蕴深厚,社会生活丰富。在调研中,我们需要运用观察的技巧,并且进行视频记录,以便分析和理解一个并不熟悉的地区。

The project will give you an opportunity to explore a culturally and socially diverse area of London. It will encourage you to use skills of observation and visual recording in the site investigation so an unfamiliar place will be analysed and understand well.

第一阶段 —— 泰晤士河南岸地区调研
Stage 1—Observation at South Bank of Thames

任务简介

分成三个小组，并对以下三个南岸的区域进行调研和分析：
Three groups explored and analysed the following three sections of the South Bank:

第一组：威斯敏斯特桥到滑铁卢桥
重要节点：郡议会大楼，伦敦眼，皇家节日音乐厅

第二组：滑铁卢桥到布莱克弗赖尔桥
重要节点：英国电影学院，皇家国家剧院

第三组：布莱克弗赖尔桥到南华克桥
重要节点：泰特现代艺术馆，千禧桥，莎士比亚环球剧院

Group 1 – Westminster Bridge to Waterloo Bridge
County Hall, London Eye, Royal Festival Hall

Group 2 – Waterloo Bridge to Blackfriers Bridge
British Film Institute, Royal National Theatre

Group 3 – Blackfriers Bridge to Southwark Bridge
Tate Modern, Millennium Bridge, Shakespeare's Globe Theatre

调研要求

首先，应当沿河步行穿过南岸，从威斯敏斯特桥步行到南华克桥。这样有助于将南岸作为一个整体进行思考，理解南岸与伦敦城及北岸的关系以及自己所调研的区域和另两个区域的关系。

First walking along the river from Westminster Bridge to Southwark Bridge to understand how the South Bank as a whole, also the relationship with London city and the north band, furthermore the relationship between our site and the other two.

之后，细致调研自己的区域，应当对以下内容着重考虑：

Then explore your section of the South Bank in more detail, paying particular regard as below:

桥
跨越河道的桥梁是如何定义区域的——谁使用它们？它们如何与河岸相连？
节点
进入和过渡的关键节点在哪儿？它们是如何被定义以及被体验的？
地标和风景
有识别性的关键风景和地标——沿河以及穿越河道，近景以及远景，好的或者坏的。
与河流的联系
你在不同的地点是如何与河流发生联系的？你可以靠近水面么？
行人和竖向交通
人们进入以及在区域内的运动行为——不仅是路线，还有运动速度以及目的地。
占有方式以及用途
谁在使用空间——当地居民、办公室职员、游客。他们在做什么？
环境参数
当地气候状况——阳光和阴影，朝向和遮阳，噪声。
建筑形式
建筑的形式、尺度、年代、功能以及材料和构造形式的混合运用。
未来发展
这个区域正在经历变化么？有没有正在建设的新建筑？这些将如何影响人们对空间的体验？

Bridges
How do the river-crossings define the area – who uses them? How do they meet the river bank?
Thresholds
Where are the key points of entry and transition? How are they defined and experienced?
Landmarks and views
Identify key views and landmarks – along and across the river, near and far, good and bad.
Connection with river
How connected to the river in different locations? Can you get close to the water?
Pedestrian + vehicular movements
Movement to, into and at the area – not just route, but speed and purpose.
Patterns of occupation and use
Who uses the spaces? – local residents, officers, tourists. What are they doing?
Environmental factors
Local climatic conditions – sunlight and shade, exposure and shelter, noise.
Building form
Type, scale, age, function and mix of buildings – materials used and forms of construction.
Future development
Is the area undergoing change? Are new buildings under construction? How might this affect experienced in this place?

运用草图、地图、图标以及照片来记录自己的观察。
要既客观又主观，这意味着应该依据记录进行分析，并且透过表面来揭示未预料的事实。个人的结论应当是个人主观的，并应反映自己的观点。作出评价——喜欢什么？不喜欢什么？这里与你熟悉的地方如何进行比较？

Use sketches, maps, diagrams and photographs to record your observations.
Be both objective and subjective – this means analysing as well as recording, and looking beyond the obvious to reveal the unexpected. Your response should be individual and reflect your opinion. Make value judgements – what things do you like, and what don't you like? How does it compare to places you are familiar with?

第二阶段——汇报
Stage 2 —Presentation in Sheffield

在谢菲尔德，将会有专门的时间整理编辑调研报告，并进行汇报。
In Sheffield, there will be a part of time to edit and collate the data for the investigation, and then for a presentation.

第一组
飞鸟雨篷

从布莱克弗赖尔桥到南华克桥
节点：泰特现代艺术馆，千禧桥
谭佳俊　张天硕　冯　喆　邵菁菁

GROUP 1
BIRD SHELTER

From Blackfriers Bridge to Southwark Bridge
Nodal point: Tate Modern, Millennium Footbridge
Tan Jiajun, Zhang Tianshuo, Feng Zhe, Shao Jingjing

沿岸状况及重要节点调研

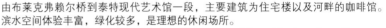

由布莱克弗赖尔桥到泰特现代艺术馆一段，主要建筑为住宅楼以及河畔的咖啡馆。滨水空间体验丰富，绿化较多，是理想的休闲场所。
From Blackfriers Bridge to Tate Modern, there are mainly residents and a coffee near the river. The waterside space can be experienced much as a perfect relax zoon.

重要节点：泰特现代艺术馆及千禧桥。这里是整个区域的高潮，作为南岸现代艺术的最中心地带，是空间尺度最大并且人群集散量最大的地方。
Thresholds: Tate Modern and Millennium Footbridge, which are regarded as the climax of the South Bank and the central area of the modern art. The scale of the space here is huge, and serve for great numbers of people.

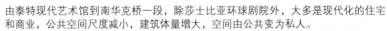

由泰特现代艺术馆到南华克桥一段，除莎士比亚环球剧院外，大多是现代化的住宅和商业，公共空间尺度减小，建筑体量增大，空间由公共变为私人。
From Tate Modern to Southwark Brige, except for the Shakespear's Globe Theater, most of the buildings are modern commercials and residents. The scale of public space shrinks, while the scales of buildings increase.

人群行为调研

一是在泰晤士河南岸虽然建筑众多,然而公共设施,特别是雨篷十分匮乏。加之伦敦天气多变,市民只得在建筑的屋檐下避雨,十分不便。
First of all, there are many buildings at the South Bank of Thames, as for the public facility, the rain shelter, is specially rare. This is realy inconvenient, when considering the weather in London, that people have to get rid of the rain under the roof.

二是虽然南岸拥有泰特现代艺术馆、莎士比亚环球剧院、国家画廊等文化建筑,然而这些艺术圣地对于城市的辐射力并非理想中的那样,所以我们打算借助雨篷,为伦敦街头艺人提供一个集遮雨与临时舞台于一身的公共设施。
Secondly, indeed there're Tate Modern, Shakespear's Globe Theater, National Gallary and many other art buildings in south bank, but these art santuraies actually have less influence to the city. So we decide to build new rain shelters should combines the function itself and a temporary stage for the artist on the street.

我们的提案是一个飞鸟造型的雨篷。之所以用鸟的造型设计雨篷,是因为泰晤士河南岸有很多鸟。鸟飞翔的动态是艺术与文化自由最好的形象化符号,所以我们要打造的是伦敦的艺术之翼,以公共艺术博物馆为中心,使艺术的氛围遍及南岸的大街小巷。

Our proposal is a rain shelter. The reason we use bird as our concept is simply that there're many bird at South Bank. The flying posture of the bird stands for the freedom of the art and culture. So we want to create a Wing of Art, which spread the atomsphere of art from museums to every where along the street.

左图为设计概念图,右图为方案模型
Concept design in left, model in right

雨篷收起时，可以作为路灯以及装饰；当艺人将雨篷打开，并滑动到最下端时，雨篷成为一个临时舞台；下雨时，雨篷可以滑动到任意高度，作为避雨设施使用；夜晚如果将雨篷打开，通过漫反射，固定在悬臂上的发光系统可以为城市提供绚丽的照明。这就是我们对于调研中问题的回应——飞鸟雨篷。

When the shelter close, it could be a street lamp; When the shelter opening and slide to the bottom, it become a temporary stage; the shelter can slide to any height to get rid of the rain When the raining day, through the diffuse reflection, the opening shelter can provide colorful lightening in the night. This is our proposal, a bird shelter.

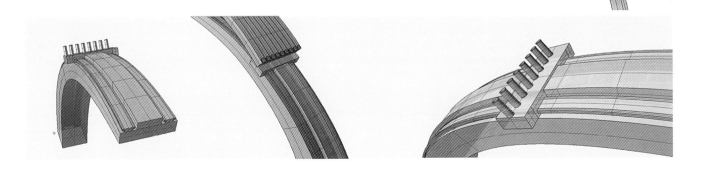

第二组
墙

从威斯敏斯特桥到滑铁卢桥
节点：伦敦眼
菅 睿　王静辉　左 煜　王翘楚

GROUP 2
THE WALL

From Westminster Bridge to Waterloo Bridge
Nodal point: London Eye
Jian Rui, Wang Jinghui, Zuo Yu, Wang Qiaochu

基地情况

基地位于泰晤士河南岸，威斯敏斯特桥到滑铁卢桥。其中主要有四座建筑，他们分别是市政厅、伦敦眼、皇家节日音乐厅和伊丽莎白女王大厅。这些建筑大部分以暖色调为主，庄重典雅，但形体上不失活泼。中间夹杂着一些现代建筑，绚丽的色彩使人眼前一亮。

The Site is located at the South Bank of London, stretching from Westminster Bridge to Waterloo Bridge. There're four main thresholds at the site, which are County Hall, the London Eye, the Royal Festival Hall and the Queen Elizabeth Hall. Most of the buildings are decorated with warm colour, making them pretty elegant. However the form is quite stimulating. Between the old buildings are some modern buildings, which inspire people by their gorgeous colour.

基地重要节点调研

Country Hall
现在是海洋馆、博物馆、酒店、饭店等的综合体。
Now it is used as complex including a aquarium, museum, a hotel and many restaurants.

London Eye
世界最大的摩天轮，伦敦的标志之一，游客云集。
The biggest ferry wheel in the world, a symble of London and a tourist attraction.

Royal Festival Hall
许多世界经典音乐在这里上演，但前面变成了餐饮中心。
Many classic music had played here, however, the front of the building become restaurants.

Queen Elizabeth Hall
比皇家节日音乐厅更加私密，经常有私人聚会。
It is more personal than Royal Festival Hall. Private parties often be held in it.

节点对比

1. 伦敦眼前场地
地处三个方向人流的交汇点，同时还是观赏伦敦眼和上伦敦眼排队等候的地方，因此这里的人流显得杂乱、拥挤。显而易见，这里非但没有什么疏导人流的设施，还有随意设置的餐饮摊点阻塞交通。雨天时更是没有任何遮雨的设施，排队的人们无奈地在雨中等待，这在多雨的伦敦并不少见。

1. Site located at London Eye
The crossing of circulation from three orientations, where is also a place for seeing and queuing to London Eye. It's crowded and also lack of guide facilities and rain shelters. The pedlar make it worse.

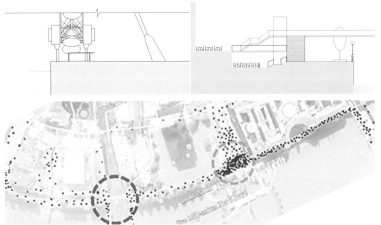

2. 铁路桥与南岸的交汇点
同样是人流的交汇点，但因为这里的空间层次划分十分多样化，将人群很有条理地疏散开，让他们走向不同的目的地，完成不同的动作，一切显得那么有序。同时，中间的台阶还提供了人们休息的场所，而错层等手法产生的灰空间给了人们避雨的场所，显得非常人性化。

2. Crossing of bridge and bank
The crossing of circulation but people are seperated regularly by the diversified layers of spaces. Furthermore, the stages provide a rest place for people, while the split-levels block layout serve as a rain shelter.

County Hall前面的街道拥挤、无层次的划分，显得混乱，尤其是摊点的桌子占了很大的面积。

通往伦敦眼的街道很宽敞，有一定层次的划分，并形成了景观，但对疏导人流作用不大。

节日大厅前空间层次丰富，空间利用率高，人流显得很有秩序，并感到愉悦。

在伊丽莎白女王大厅前形成错落的空间和檐下灰空间，成为了涂鸦和滑板的聚集地，同时设施也有趣味。

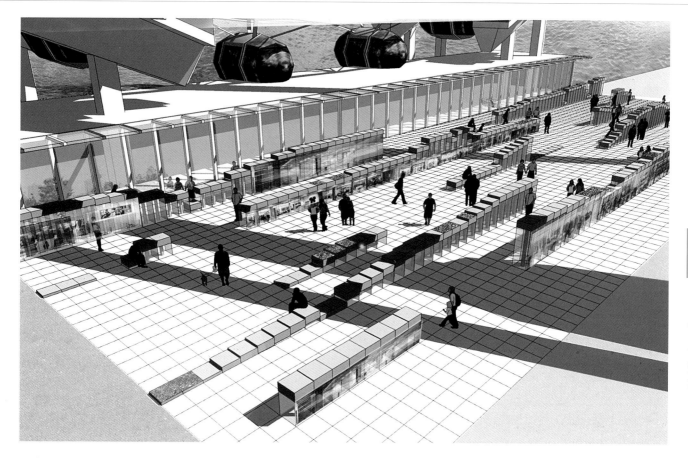

方案

方案想要将伦敦眼地区的交通通过有序的疏导,达到交通流线明晰、互不干扰又互相渗透的效果,借助的要素就是墙。通过不同高度的墙产生不同的隔断效果,形成不同的空间感受。与此同时,借助不同的高度,设置不同的功能,让空间更加富有多样性和趣味性。

细节

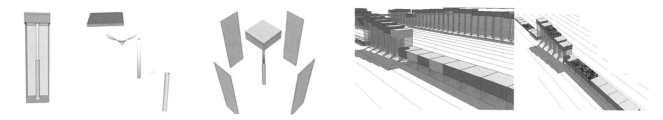

结构的升降采用液压系统,用它将结构送出地面,到达指定的高度。同时,它是整体的支撑结构,保持结构的牢固和稳定。周边的玻璃上带有LED显示屏,可以支持视频播放功能。顶盖中可以伸出挡板,作为多雨的伦敦环境中的一处遮雨设施。

第三组
伦敦泡泡

从滑铁卢桥到布莱克弗赖尔桥
节点：英国电影学院，皇家国家剧院
常湘珩　孙宇璇　钟兰婷　徐晓铎

GROUP 3
LONDON BUBBLE

From Waterloo Bridge to Blackfriers Bridge
Nodal point: British Film Institute, Royal National Theatre
Chang Xiangzhou, Sun Yuxuan, Zhong Lanting, Xu Xiaoduo

1. 功能性的需求已经实现
2. 公共空间乏味
3. 河岸与河流的互动不足

1. The functional demand is fulfilled
2. The public space is tedious
3. The interaction of the bank and the river is deficient

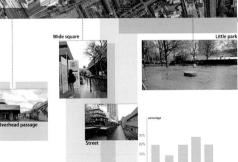

节点分析以及空间尺度分析
Thresholds and Spacial Scales analysis

内部空间分析

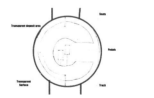
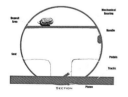

总平面图

04

6 Degrees Celsius

— The Future of A Residential District in Harbin

Chinese Teachers :	Lecturer Dong Yu	董 宇　讲师
	Lecturer Liang Jing	梁 静　讲师
	Lecturer Shi Ligang	史立刚　讲师
Foreign Teacher :	Prof. Irena Bauman	伊莱娜·鲍曼　教授

Students :	Wang Jinghui	王静辉
	Xie Aiwen	谢媛雯
	Feng Zhe	冯 喆
	Hu Xiaoting	胡晓婷
	Zhang Huiting	张荟亭
	Chen Xingyue	陈星月
	Zhang Daiyan	张黛妍
	Du Pengfei	杜鹏飞
	Zhao Yingqi	赵英亓
	Xu Xiaoyu	徐晓宇
	Song Minqi	宋敏琦
	Wang Yajun	王亚隽
	Guo Qishen	郭起燊
	Wan Shaoran	万邵然

Academic activity : International Collaborative Teaching and Academic Month
Institutions : Harbin Institute of Technology and The University of Sheffield

2012 Autumn

"6"度
——关乎哈尔滨未来的设计

任务背景

未来的50年内,全世界将经历6℃的气温提升以及随之而来的降雨量增加和更多的极端天气。与此同时,世界人口将会超过90亿。城市将不得不应对扩张与萎缩,粮食、水资源的短缺,洪涝灾害,环境难民的持续涌入以及许多无法预测,却不断发生的其他变化。以上问题是人类与城市必须适应和面对的,而解决方案的核心就是通过设计来实现。

In the next 50 years, the period in which you are hoping to work as an architect, we will experience up to 6 degrees increase in temperature, increase in rainfall and increase in extreme weather events. In the same period, world population will exceed 9 Billion. Cities will have to adapt rapidly: we will experience sprawl and shrinkage, food shortages, water shortages, floods, constant influx of environmental refugees and many other changes but most importantly constant uncertainty. People and cities have to adapt and design is central to the solution required.

作为未来的建筑师,想要在环境和可持续领域做出贡献,就需要培养自身在动态、变化的环境中的设计能力以及在一个综合的团队里的协作能力。另外,还要为他人的提升与价值实现创造条件,保持持续的研究与知识更新,更要把握住未来的发展趋势。其中的重中之重,是创造新的建筑类型;改造既有建筑,使其适应新的生活机制与生活方式。

In order to make a significant contribution you, as an architect, will have to develop skills to work in a dynamic, ever changing environment, work collaboratively in large and complex teams, learn to facilitate others to help themselves, constantly research and update your knowledge and be able to scan future trends. Above all you will need to design new building typologies and adapt the existing ones to suit the new institutions and new ways of life .

基于此,本设计训练旨在为2050年的居民社区提供一体化的设计策略。

我们将:
· 研究未来气候的状况及其对居住密度、施工方法、材料以及生活方式的影响;
· 提出地段的整合设计策略;
· 设计出新的住房类型以及新的社区机制。

我们还将:
· 协同开展设计工作;
· 了解团队工作以及个人角色。

This projects aims to develop design for a neighborhood for 2050.

We will:
Explore future climate scenarios, and the implications on densities, construction methods, materials and life style.
Develop a joint strategy for the site.
Develop a design for new housing typologies and new neighborhood institutions.

We will also:
Work collaboratively.
Learn about how team work and what role you play within it.

合作背景

谢菲尔德大学建筑学院以其"技术研究型设计教学"在欧洲乃至世界范围内的建筑类院校中闻名，而哈工大建筑学院亦在尝试着从"项目型设计教学"向"研究型设计教学"转型，因此两院的联合学术合作及教学实验有着长期的建设性意义，对哈工大建筑学院在全国范围内确立自身的办学特色和教学特色也有着非凡的积极意义。

Sheffield School of Architecture have been well-known for its "technical research-based teaching" among schools of architecture not only in Europe but all over the world. In the meanwhile, HIT School of Architecture has been trying to convert its teaching pattern from the "Project-based teaching" to "research-based teaching". Therefore, the academic cooperation and teaching experiments between the two schools have far-reaching significance, including the positive significance that HIT can establish its own characteristics nationwide on both the way of running a school and the way of teaching.

作为"6度未来设计"的中方合作教师，我们在与Irena教授的共事中，除了感受到她充满活力的教学风格与教学魅力以外，也深受谢菲尔德大学成熟的教学模式的启发。在课程启动前期，我们与Irena教授保持着一个多月的通信往来，通过电子邮件交流教学方法，提供学生的相关信息，使其了解学生的设计水平与设计表达能力，并利用google earth等地理信息软件提取坐标及基地尺寸，来商讨地段选取标准，并最终确定设计地段。

As Chinese co-teachers working with Professor Irena in Project "six degrees future design", we are impressed with her passionate teaching style, at the same time we are deeply inspired by the mature teaching pattern of Sheffield. Before the course started, we had been contacting Prof. Irena by e-mails for more than one month. Through e-mails, we shared ideas about teaching methods with Prof. Irena, and provided with relevant information about the students, to inform her the students' designing and expressing ability. We also extracted coordinates and size of the site with Google Earth and the other geographic information software, to discuss standards and the final location for the site.

全体同学和老师们的合影
The photo of all the students and teachers

DAY	TITLE	KEY TASKS	OUTPUTS	YOU WILL NEED
Pre-summer school		TASK 1: Research how climate change will impact on Harbin	1 A4 side summary of your research	
		TASK 2: Think of a small object that means somethng to you.	You will bring this in on Monday to show to the others.	
Monday	Getting to know each other and the project	TASK 1: Present your small object that means something to you to the others	Getting to know each other	• A small personal object • Thick pens • Measuring tape, ruler etc • Camera • A3 paper • Sketchbook • Scalpel/craft knife • Glue • Masking tape • Simple modelling materials (for site model)
		TASK 2: Site visit and analysis Start conversations with local people	Gain understanding of what the site is like and who lives and works there	
		TASK 3: Build a working model of the site	Group site model to use throughout the week	
Tuesday	Scenario for 2080 - What kind of neighbourhood?	TASK 1: Review of site analysis and conversations with local people	Gain a deeper understanding of how the site works and what is needed in the area	• Original site work • Print out A4 portraits of the local people you met • 1 A4 side summary of research (Pre summer school Task 1) • Thick pens • Sketchbook
		TASK 2: Discuss in small groups your research on how climate change will impact Harbin	Agree what key climate adaptations are needed in Harbin	
		TASK 3: Present your ideas to the whole group and brainstorm a brief for the site	A group brief and decide who will work on each adapation	
Wednesday	Develop initial proposals	TASK 1: Develop outline design for your project	Development massing models, rough drawings and layout	• Modelling equipment and materials • Drawing equipment and materials
		TASK 2: Make a simple sketch model of your proposal	Everyone to insert sketch models into site model to gain an overview of neighbourhood	
		TASK 3: Reflection on your proposal so far	Prepare for 10 minute presentation of your proposal on Thursday	

DAY	TITLE	KEY TASKS	OUTPUTS	YOU WILL NEED
Thursday	Neighbourhood taking shape - does it work?	TASK 1: 10 minute presentation each to rest of the group	Review and discuss each other's proposals	• Modelling equipment and materials • Drawing equipment and materials
		TASK 2: Continue development on reflection of comments made	Development of individual proposal	
		TASK 3: Prepare a group presentation and short individual presentation (Total 30 mins for both)	Presentation to tutors/local residents tomorrow morning (Friday)	
Friday	Testing ideas with other people	TASK 1: 30 minute group and individual presentations	Gain feedback from tutors and residents	• Modelling equipment and materials • Drawing equipment and materials
		TASK 2: Reflect on feedback and develop final drawings for your proposal	A few drawings that help explain the key ideas behind your proposal	
		TASK 3: Make a final model of your proposal	Your final proposal to insert into the site model	
Saturday	Bringing it all together	TASK 1: Continue Tasks 2 & 3 from Friday	Key drawings and final model	• Modelling equipment and materials • Drawing equipment and materials
		TASK 2: Prepare a collective presentation of the neighbourhood plus individual proposals for a final review tomorrow (Sunday)	Presentation ready for final review tomorrow (Sunday)	
Sunday	Final presentations	TASK 1: Group and individual presentations		• Final drawings and model • Camera
		TASK 2: Enjoy and take credit for all the work!		

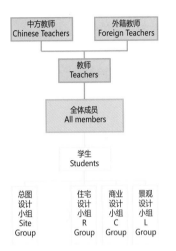

团队组织框架图
The framework of the team organization

团队组织框架

14人的团队做一个项目，这中间的合作、分工、协调是一个非常有意思的事情。小队的合作，小队与小队之间的协调，大家共同的讨论甚至争论，个人的退让与妥协，我们经历了这些，学到了很多，也理解了很多。这将是一次难忘的经历。

Team of 14 students will do one project, the cooperation, collaboration and coordination is very interesting. Cooperation in a team, coordination between different teams, discussion and debates among each other, and compromise of oneself, which we have experienced has teached us a lot. This is going to be one of our unforgettable experiences.

团队协作流程

设计时，首先是全体成员一起讨论，将重要的东西记录下来，然后由总图组整合大家的意见，经过设计，将总体的设计思路确定下来，继而将总图组的成员分到商业、景观、住宅等各个组中，作为小组的骨干力量进行分组设计，控制设计的大体方向与风格，落实总图中的设计意图，然后通过模型整合各组的成果，大家共同讨论，找出问题，修改设计方案。在设计的过程中，各个小组之间的协调统一主要通过总图组成员之间的讨论和协商解决问题。期间，更是邀请景观设计的老师及其团队来为我们的设计作指导，探讨景观组设计的可行性。总之，整个过程中，沟通和协调是关键，如何让各个小组的设计成果在组合之后更像是一个设计，需要许多轮的探讨和修改。

First, all members join in to discuss and record important things are needed. Then Site Group will integrate everyone's opinion, and give general design thoughts. Next, members from Site Group are assigned to Commercial Group, Landscape Group, or Residence Group as the main designers to control the general design style and carry out the original thoughts. Then, integrating the results of all the groups through models, all teams discuss together, find problems, and fix them. During the design process, the coherence between various groups mainly depend on consultation of the members of Site Group. During that time, we have also invited a teacher of landscape design with her team to guide us about the design, and explore the feasibility of Landscape Group. Anyway, communication and consultation are the key to the whole design process, and more rounds of discussion and modification are needed to make sure the design results of each group after combination is still one design.

激烈的讨论

给前来支援的景观老师讲述方案

商业小组制作大比例模型

总图小组的方案推敲模型

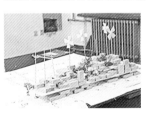
商业小组的大比例模型

各组体量整合模型

王静辉
Wang Jinghui

陈星月
Chen Xingyue

谢嫒雯
Xie Aiwen

张荟亭
Zhang Huiting

冯喆
Feng Zhe

王亚隽
Wang Yajun

胡晓婷
Hu Xiaoting

宋敏琦
Song Minqi

万邵然
Wan Shaoran

徐晓宇
Xu Xiaoyu

张黛妍
Zhang Daiyan

杜鹏飞
Du Pengfei

郭起燊
Guo Qishen

赵英亓
Zhao Yingqi

社区方案的最终成果模型，材质为雪弗板，尺寸为2520mm×1500mm×400mm。
The photo of final model of the design material：Chevron Board，size：2520mm×1500mm×400mm.

基地调研

首先进行的是基地的调研。该调研分成三组进行，通过实地考察、亲身体验、现场访问等手法了解基地的信息。对基地中的建筑进行区分，通过对这些建筑的了解，确定哪些是该保留的，哪些是需要拆除的。

First is survey of the site. During the survey, all the students are divided into three groups, visiting the site, experiencing by themselves, and interviewing local residents to get information about the site. Then classify buildings in the site through the understanding of them, and determine which are to be the reserved, and which are to be demolished.

访问基地中的住户

老师的集体讲解

特色空间——作坊式的铁艺工匠铺

针对调研制作模型，讨论建筑的保留

标志性构筑物——烟囱

临街住宅入口空间

楼间花坛的空间处理

街头转角处的公共厕所

基地分析与方案生成

基地周边中心绿地
Green land around the site

包围基地的商服带
Commercial district surrounding the site

提取基地内部与周边的三个烟囱作为制高点
Choose the three chimneys as commanding heights

连接绿地与制高点形成中心景观"绿线"
Connecting greenland and commanding height as landscape green line

烟囱成为"自行车站"和至高观景平台
Chimneys as bicycle stop and viewing platform

基地原有建筑纹理为横向，均分基地为四部分
Original pattern divided the site into four equal parts

将基地划分为"公共"与"私密"两个分区
Separate the site into public and private districts

根据开放度分为四个等级
Arrange the site into 4 parts according to openness

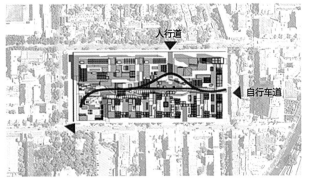

人行道
自行车道

SOHO 区
SOHO Quarter

商业区
Business Quarter

台阶景观
Step Landscape

保护建筑广场
Protected Building Square

自行车专道
Bicycle Lanes

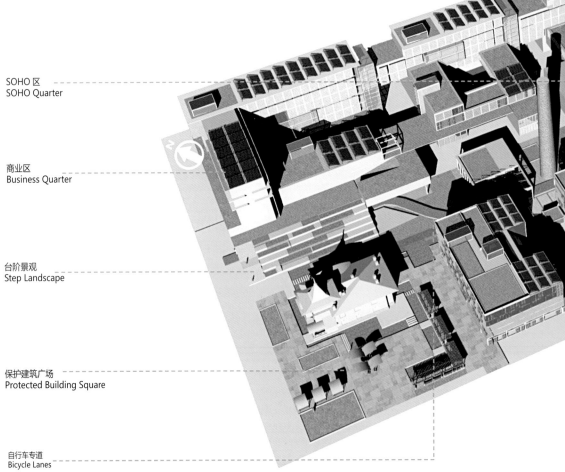

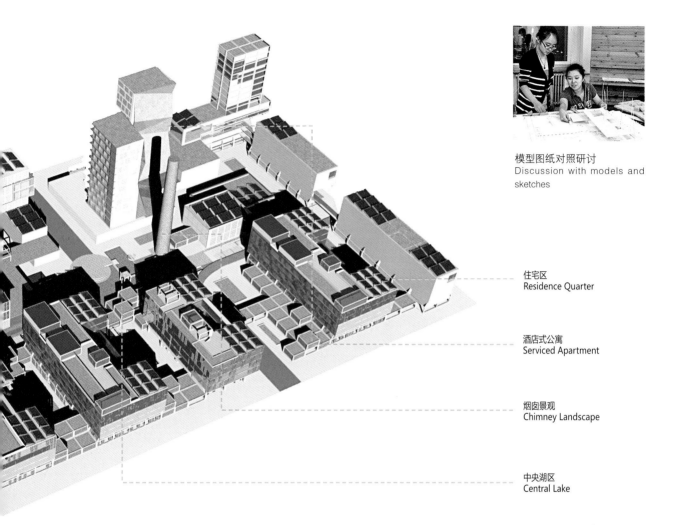

模型图纸对照研讨
Discussion with models and sketches

住宅区
Residence Quarter

酒店式公寓
Serviced Apartment

烟囱景观
Chimney Landscape

中央湖区
Central Lake

规划设计

整体布局以三个烟囱为制高点，以连接绿地广场和马家沟河两个自然景观要素的一条节能"绿线"为主轴线，烟囱除了其自身功能外，更承载了作为"自行车站"和观景平台的新功能。因为烟囱的高度使其在远距离更易被发现，可以成为骑行的地标，将城市中的烟囱改造为"自行车网点"，结合补给、商服等功能，鼓励更多人骑车出行，打造环保新社区。

The overall layout take the three chimneys as commanding heights, and take the energy saving "green line" which connects the two natural landscape-the green square and Majiagou river as main axis. In addition to its own function, the chimneys are given new functions as "bicycle stops" and landscape platform. Because of the height of the chimney, it can be easily found in the distance so as to be a landmark of cycling. Then change the chimneys in the city into "bicycle network", combine it with the supply and commercial functions, encourage more people to take riding as a common transportation, and create a new environmental-friendly community.

住宅可变性

我们考虑的出发点为可变的、持续的、舒适的、田园化的、与环境友好的等,这些出发点在我们的建筑形式、结构形式、建筑立体绿化、建筑表皮、利用可持续清洁能源等方面,都有所考虑和体现,最终的考虑结果如下所述。

根据人的不同需求设置不同的套型面积、套型样式,并可进行变换,一切随心所欲。

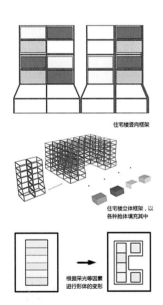

集合住宅部分是由框架和舱体组成的可活动住宅,且为了实现住户多面采光,单体框架组成"U"形框架作为一个建筑单体。

住宅的基本单位——舱体可以进行单个或多个的灵活组合以满足不同住户的不同需求,如大家庭、单身住户,还有老人。我们设想了一些房间组合,当然,这都可以根据住户的要求进行调整,而暂时空出的框架格则作为立体绿化的公共空间。当然,有固定位置的框架作为交通盒并排布管线。

空间与节能

3~6层为居住空间，1~2层是公共空间，可以进行灵活的调整利用，可作为停车空间、商铺空间以及供人们休憩的半室外空间。沿街部分通过几个舱体相连，相连的部分作为商铺或进入小区的入口以及小区的公共空间，并增强立面的完整性。

建筑顶层以及空置的框架格中考虑了立体绿化的设置，并且在顶层设置了太阳能电池板，为建筑提供一部分的能量。在建筑的外部，利用可活动的自动调节百叶进行围护，减少热交换，为建筑节省能源，创造更好的建筑内部小气候。

公共空间
Public Space

商业空间
Commercial Space

一、二两层的小区停车场

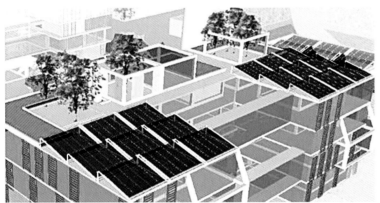

屋面太阳能光伏板为住宅提供能量，立面上的百叶用来调节舱体微气候

楼与楼之间通过体块相连，构筑街道立面的完整性

商业空间形态与布局

该区域建筑以多层的SOHO商居一体化住宅为主体，辅以其他灵活的各种功能空间，创造出全新的多功能半开放式综合社区。

原有的沿街道的市场被整合到建筑内部，建筑一层大部分空间以及部分二层空间被用作新型半开放交易市场，释放出沿街空间，更加畅通美观。面向小区内部的一侧采取了开放式处理，一方面呼应了对面普通住宅区的开放性，一方面营造出了良好的社区氛围。我们结合核心景观设置了多个半开放的院落，供居民休闲停留，建筑内界面沿着景观灵活退让，层次丰富。

高度上与周边其他原有住宅建筑齐平，获得天际线的统一；同时，临街界面在平面上也与原有界面齐平，保持原有肌理。这样的统一保持了大体上的和谐。同时，在转角处放置了两幢高层建筑，提高了土地使用效率，高层部分的技术应用极大地降低了能耗。

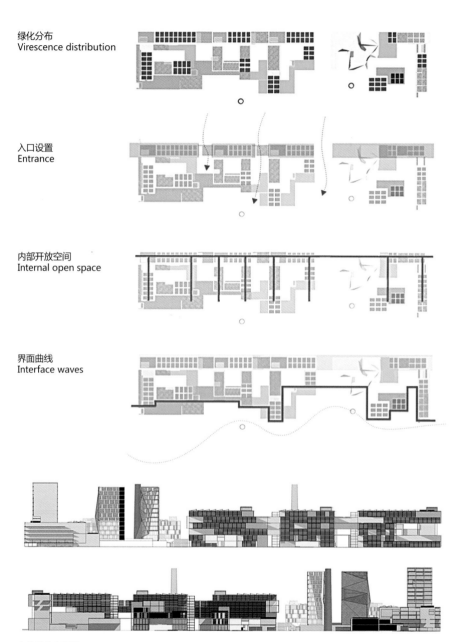

绿化分布
Virescence distribution

入口设置
Entrance

内部开放空间
Internal open space

界面曲线
Interface waves

商业部分立面图
Commercial part facade

生态节能措施

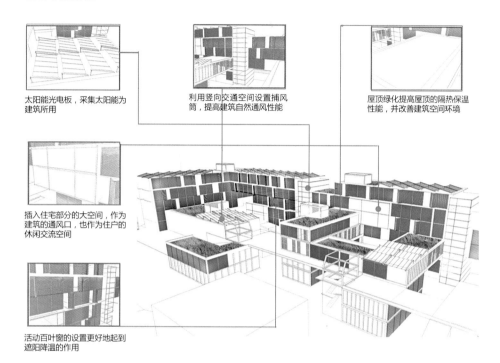

太阳能光电板，采集太阳能为建筑所用

利用竖向交通空间设置捕风筒，提高建筑自然通风性能

屋顶绿化提高屋顶的隔热保温性能，并改善建筑空间环境

插入住宅部分的大空间，作为建筑的通风口，也作为住户的休闲交流空间

活动百叶窗的设置更好地起到遮阳降温的作用

从沿街面到内侧，建筑形体逐级跌落，形成外高内低的台阶状布局。每层屋顶结合具体情况设置绿化或活动场地以及太阳能集热板，这样的设计合理利用了资源，提高了绿化面积，丰富了景观层次，扩大了采光面积，同时为整个小区的生态循环系统提供了更方便有效的平台。

The shape of the building fall from along the street to the inside step by step, forming a layout of step-shape with higher inside and lower outside. Each layer is set with green roof or activity place and solar panels, according to the specific circumstances. Such design use resources wisely, increases the green area, riches landscape layers, expands the lighting area, at the same time provides more convenient and effective platform for ecological cycle system of the whole community.

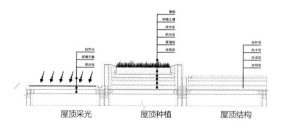

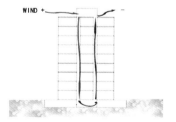

屋顶绿化不仅仅有保温隔热性能，还能有效改善空气质量，隔绝噪声
Green roof not only have heat preservation and heat insulation performance, but also can effectively improve air quality and isolate noise.

捕风筒加强建筑内部空气流动，改善居住环境
Air-catching cylinder strengthens internal air flow, and improves the living environment.

现实中的梯田景观
The terraced landscape in reality

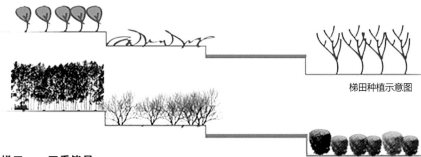

梯田种植示意图

梯田——四季皆景

沿着地势的高低种植不同高度的植被，并按照季节的不同进行精心搭配，使居住区内时时有景，处处成趣。这些梯田也是水循环系统中的重要一环，可以储水、净水、回收水，成为了小区内的核心景观带。大台阶位于SOHO片区的端部，住宅区里的人们可以直接沿台阶而上到达二层室外休闲绿台，也可以从台阶下穿越，进入下一段道路。

Along the height of terrain, plant different height of vegetation, and change the collocation according to the season, to create scenery as well as fun in the community. These terraces are also an important part of water circulation system, which can save water, purify water, recycle water, and serve as central landscape inside the community. The landings for landscape are located at the end of the SOHO area, people in the neighborhood can directly reach green platform on the second floor by following the steps, or go to next road down through the steps.

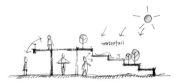

台阶景观剖面示意图
Sketch of section of the landings for landscape

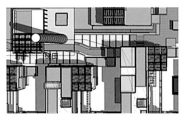

梯田景观平面图

保护建筑广场——邂逅古今

古建广场的处理保留了原有的韵味，古朴大气，意味深远。居住区内留出宽敞的空间作为广场，使其既可以展露于城市界面，又可以作为居民茶余饭后的休闲场所，水域与长廊的萦绕更增添了空间的趣味。

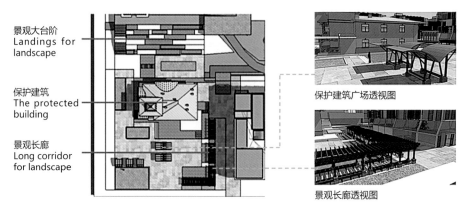

景观大台阶
Landings for landscape

保护建筑
The protected building

景观长廊
Long corridor for landscape

保护建筑广场透视图

景观长廊透视图

中央湖区

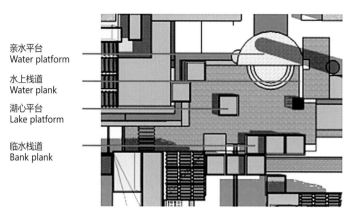

亲水平台
Water platform

水上栈道
Water plank

湖心平台
Lake platform

临水栈道
Bank plank

景观设计以水系为中央主轴，而中央湖区是其核心。整个中央湖区作为住宅和商业部分的衔接，风格与前两者相统一。手法上着重使用统一的形体元素，方体元素的湖心平台，方体元素的临水步道，方体与住宅以及商业区的构成元素一致。

中央湖区夏季为居民提供亲水场所，冬季则会成为小区居民滑冰的好场所，而户型平台则可以成为他们滑冰之余休息娱乐的场所。

烟囱景观

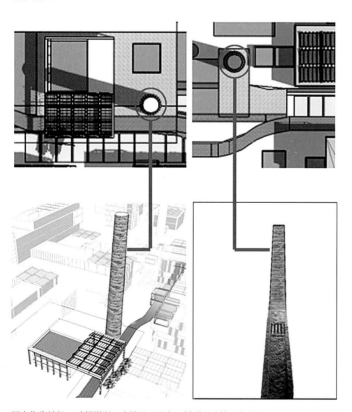

烟囱高高耸立，这是我们的设计主要保留的构筑物之一，因此发掘烟囱的各种功能以及景观效应成为了我们的方向。首先，这里是自行车骑行的中转站，提供休憩、餐饮、观景空间。同时因其高度，它也成了骑行的路标。另外，作为城市的瞭望塔，在这里举目远眺，周边的风景尽收眼底。

我们让攀爬的植物顺势而上，带给它历史和沧桑的感觉，同时也让它更加具有观赏价值。

烟囱作为地标、瞭望塔以及立体景观平台；其附属建筑具有休憩、餐饮及观景功能

05

Afternoon Tea

— Academic Activity: International Collaborative Teaching and Acacemic Month

Chinese Teachers : Lecturer Lian Fei　　　　连　菲　讲师
　　　　　　　　Lecturer Meng Qi　　　　 孟　琪　讲师
Foreign Teacher : Prof. Isabel Maria Britch　Isabel Maria Britch　教授

Students :　　Huang Yumeng　　黄雨萌
　　　　　　Shi Yuchi　　　　　石宇驰
　　　　　　Guo Qianli　　　　 郭乾立
　　　　　　Wang Yinuo　　　　王一诺
　　　　　　Cui Hao　　　　　 崔　浩
　　　　　　Zhang Zeyou　　　 张泽友
　　　　　　Cao Jingxian　　　曹婧娴
　　　　　　Chen Biting　　　　陈碧婷
　　　　　　Wei Guomi　　　　 韦国咪
　　　　　　Zhao Suyang　　　 赵苏扬
　　　　　　Yao Jian　　　　　 姚　健
　　　　　　Shu Shaoyin　　　 舒绍银
　　　　　　Fang Haochen　　　方浩臣
　　　　　　Zhou Shuo　　　　 周　硕

Academic activity : International Collaborative Teaching and Acacemic Month
Institutions : Harbin Institute of Technology and The University of Sheffield

2012 September

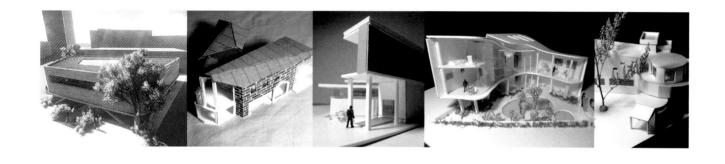

"下午茶"——学术活动

任务背景

  茶壶矛盾地蕴含着两种意境:独酌时的慰藉,共聚时的欢愉。令我们费解的是这两种矛盾的意境是如何在茶壶这一种普通生活用品上同时存在的。——佚名
  建筑无疑是一种容器。建筑内部空间之于建筑就相当于壶中香茗之于茶壶本身。我更希望人们所津津乐道的不只是茶壶本身,而更多在于壶中的香茗。——谷口吉生
  人的一生中几乎没有什么时光比投身于一种叫作"下午茶"的仪式所花去的时光更能令人愉快的了。——亨利·詹姆斯《贵妇肖像》

'strange how a teapot can represent at the same time the comforts of solitude and the pleasures of company' Unknown

'Architecture is basically a container of something. I hope they will enjoy not so much the teacup, but the tea.' Yoshio Taniguchi

'There are few hours in life more agreeable than the hour dedicated to the ceremony known as afternoon tea.' Henry James, *The portrait of a lady*.

本次设计与茶相关。茶是一种源于中国的作物,很久以前传入英国。如同品茶在中国是一种传统一样,在英国饮茶也成为了一种独特的饮食习惯。本次设计意在使两国文化传统联系、交流并探索其中的不同与相同。在曼彻斯特,我们有一座中国式茶庄,同时兼具中国艺术中心的功能。本次设计的任务是设计一栋兼具英国式茶室和英国艺术中心功能的建筑。

英国式茶室应包含以下功能及面积要求:
- 茶室(应包含小储藏室和厨房)　　　　　　100m^2
- 接待室　　　　　　　　　　　　　　　　　12m^2
- 展廊　　　　　　　　　　　　　　　　　　125m^2
- 艺术家工作空间　　　　　　　　　　　　　25m^2
- 艺术家住宿空间　　　　　　　　　　　　　25m^2
- 管理办公空间　　　　　　　　　　　　　　30m^2
- 卫生间(男、女和无障碍)　　　　　　　　 15m^2

This project is about 'tea' - a product which came from China to England many years ago and which has become typically English as well as typically Chinese. The intention is to make connections between our two countries and to explore cultural differences and similarities. In Manchester, we have a Chinese art centre and tea-shop; your task is to design an English art centre and tea rooms in Harbin.

The English Tea Rooms should include:

- Tea Rooms (including small store-room and kitchen)　100m^2
- Reception　12m^2
- Gallery　125m^2
- Artist's Project space　25m^2
- Artist's accommodation space　25m^2
- Office space　30m^2
- Toilets (male, female & disabled)　15m^2

合作背景

"下午茶"的外方教师——来自谢菲尔德大学建筑学院的Isabel Britch女士，同时也是Architects Britch Ltd.的一名职业建筑师。建筑事务所所在的办公大楼中有一间中国艺术中心，兼有中国茶室和艺术展览的功能，可以让英国人了解中国传统茶艺以及中国当代艺术和文化。本次联合设计的题目正是来源于此，Isabel提出在中国的哈尔滨设计一座展示英式下午茶文化的英国当代艺术中心来促进中英在茶文化以及由此延展而来的艺术和文化方面的了解与交流。

作为"下午茶"的中方合作教师，在与Isabel共事的一段时间里，深刻体会到了英国教学模式与中国学习模式之间的碰撞，由此获得了极大的启发。"每一步设计，即使是灵感的迸发，都要有充分、严谨的依据"，是这次联合设计中Isabel的教学带给我的最深体会，而这与她轻松、幽默的教学风格共同构成了本次联合设计的美好回忆。

Ms. Isabel Britch comes from the School of Architecture of the University of Sheffield. She is the foreign teacher of our "afternoon tea" design project and a professional architect of Architects Britch Ltd. as well. In the office building of her Ltd. locates a Chinese art centre with functions of a tea shop and art exhibition. This center makes the British get to know more about the culture of traditional Chinese tea art and modern Chinese art. Based on the facts mentioned above and the inspiration triggered by them, Isabel asked us to undertake a project that is to build a British art centre filled with British culture of afternoon tea in Harbin of China. This project is meant to facilitate the communication and comprehension in the tea art between China and Britain as well as the modern art and culture.

As a Chinese cooperation teacher, I profoundly realized the colliding of British teaching mode and Chinese learning mode during the period of time working with Isabel. "Every step of the design, even the inspiration, has to be based on sufficient and precise evidence. " That's what I learnt from Isabel's teaching in our cooperative process which combined with her humorous and lively teaching style left us so many beautiful memories.

工作日程

星期一 —— 基地调研
调研基地：观察——城市特点、交通、人流、方位、气候、噪声等。分析区位与更大范围的街区乃至整个城市的关系。
选择一条与茶有关的名句，上文中提到的或者你曾经听到过的均可。仔细品味其中含义并与本组成员们讨论。

星期二 —— 基地分析
分小组进行基地分析工作。
这需要我们把收集的所有资料在整个大组中共享。我们将把以下不同任务分配到小组中去：基地模型（1∶500）的制作、使用意向调查、周边环境调研和摄影记录。

星期三 —— 基地策略
你们对基地实施的策略都要经过课堂讨论。将你们对茶及建筑的想法进行整合，然后设计制作一个1∶500的概念模型。这个模型的尺寸要求是：能恰好放入整个大组之前制作的那个基地环境模型里。

星期四 —— 概念模型
你们对本次设计的想法都要经过课堂讨论和审查。为了彻底检验你们的想法，你们现在应当开始着手绘制1∶100的平面图和剖面图。

星期五 —— 平面和剖面
你们的平面图和剖面图以及表现草图或三维视图都要经过课堂讨论。你们现在要用1∶200附带基地周边环境的总平面图来表达你们的方案。然后你们需要绘制表现草图或制作模型来探讨你们的方案是如何构成的。

星期六 —— 最终成果
完成你们的设计方案。你们提交的最终成果应含有以下几项：
- 基地策略
- 概念模型
- 总平面图
- 平面图和剖面图
- 附带周边环境的三维表达视图
- 最终模型

星期日 —— 成果展示与成果讨论审查

Monday – Site Visit

Visit site: Observe, urban character, traffic, people flow, orientation, weather, noise etc…Analyse the sites relationship to the wider district and the city.

Choose a quotation about tea, one of the above or one which you have heard before.
Think about its meaning and discuss with the group.

Tuesday – Site Analysis

In groups carry out site analysis work
This is about gathering information to be shared by the whole group. We will allocate different tasks including site model 1:500, buildings-use survey, wider context and photographic record.

Produce an individual site strategy diagram in which you identify the key aspects or characteristics of the site and its relationship to its context.

Wednesday – Site Strategy

Your site strategy will be discussed. You will then be asked to pull together your ideas about tea and about the building and to create a concept model at 1:500. This model will fit into the overall group model.

Thursday – Concept model

Your ideas for the project will be discussed and reviewed. You should now start to draw plans and sections at 1:100 scale in order to test out your ideas.

Friday – Plans & Sections

Your plans and sections will be discussed and any sketches or 3D views. Now draw your scheme at 1:200 in its wider context and sketch or make models to explore what it may be made of.

Saturday – Completion

Complete your project. Your submission should include:

- Site strategy
- Concept model
- Plan in context
- Plans & Sections
- 3D views in context
- Final model

Sunday – Review

方案一

学生姓名：黄雨萌
哈尔滨工业大学09级本科生
The Undergraduate Designer:
Huang Yumeng

基地策略
Site Strategy

此基地位于哈尔滨西大直街，哈尔滨工业大学正门对面一个七岔路口旁。我将茶室在一层分为两部分，中间留有一条小路为行人提供方便。同时，为将行人吸引至茶室，我将整个茶室的核心部分，即开放式茶道表演厨房放在了这条小路旁。大街岔路口一侧的茶室，是以现代化为主题的茶室，而邻古典建筑一侧的茶室则保留传统英国茶室的风格。为了退让出人性化的空间作为茶室与古典建筑之间的联系，并使古典建筑成为古典茶室一个很好的借景，我在茶室与古典建筑之间建了一座小庭院。茶室的二层主要为艺术展厅，我将艺术家的居室与创作室连在一起设置在二层的中心部分。

布局生成

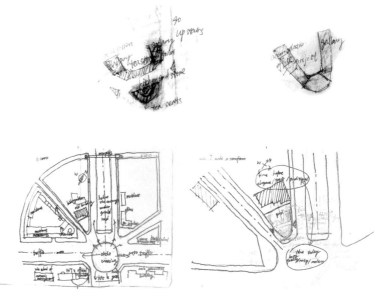

一层平面图   二层平面图

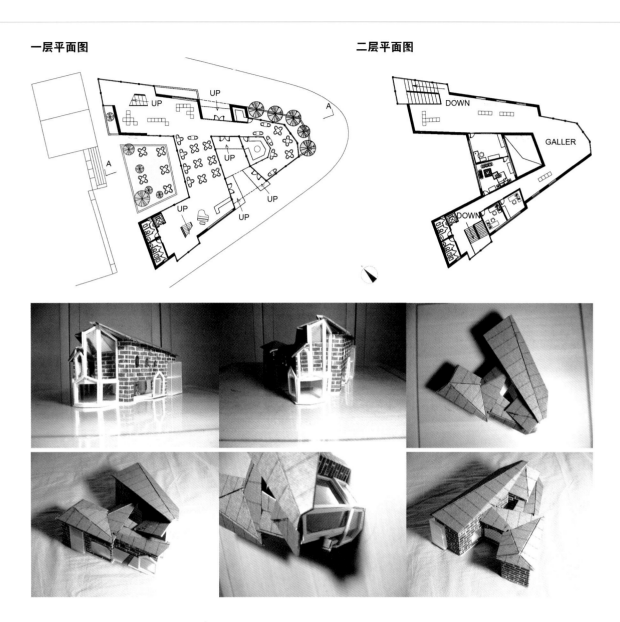

A-A 剖面图

方案二

学生姓名：石宇驰
哈尔滨工业大学09级本科生
The Undergraduate Designer:
Shi Yuchi

基地策略
Site Strategy

上班族两点一线的忙碌生活，令他们身心俱疲。所以，为繁忙的上班族创造一个静静品茶、放松冥思的庇护所成为我这次设计的首要理念。

本次基地选址在一块位于街角的三角地带上，处于南岗区最繁华喧嚣的地方。主要服务人群为附近的上班族和部分哈工大学生。沿着临街界面设计两面高墙，用来屏蔽来自外界的喧嚣，并由此形成内部的小型室外庭院，令人们在品茶休憩的同时更加贴近自然。

平面及推敲

课堂 In Class　　草模 Sketch　　基地模型 Site Model

平面图

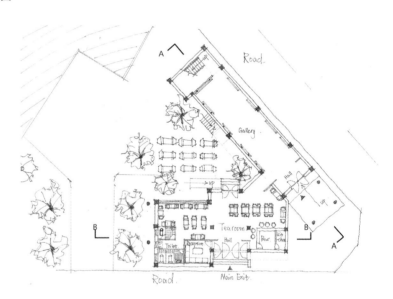

剖面图 A-A

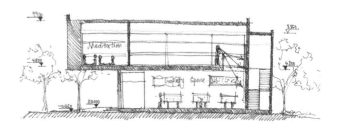

局部透视图

内部庭院处理
Interior Garden

临街立面
Street Facade

全景透视
Perspective

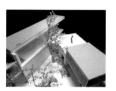
室外艺术家活动平台
Outdoor Activity Platform For Artists

室外入口
Exterior Entrance

Isabel女士说"适宜的质地及颜色较之繁复的体量也许更能直观地为人们所接受和感知。"这一点让我记忆深刻，也让我在以后的设计中会越来越多地考虑人们的使用需求及其心理感受。

Ms. Isabel once said, "the apprppriate texture and color are more acceptable and perceivable for people to understad directly than the complicated body mass." which impresses me deeply, and let me in the future consider the design of people's needs and their psychological feelings.

剖面图 B-B

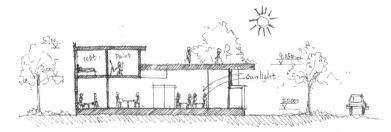

方案三

学生姓名：郭乾立
The Undergraduate Designer:
Guo Qianli

基地策略
Site Strategy

有位英国诗人Alexander Pope在1712年写的The Rape Of The Lock（《额发的凌辱》）中说道：三邦是服，大哉安妮；时而听政，时而啜茶。这句诗正描绘了英式茶文化的深邃意境：作为闲暇时光的最佳伴侣，茶的美味离不开每天的辛勤工作。

本茶室设计理念有两个：
一是英式下午茶氛围的营造。茶有助于工作，但是人们总是忙碌于工作，连一杯茶都无暇享受。我们很有必要创造一个专门的饮茶空间，远离都市的喧嚣，好好享用手中的热茶。
二是地点位于闹市与住区的过渡地带，因此茶室也应在声音、生活方式及场所感上完成过渡。

初期草模

平面图

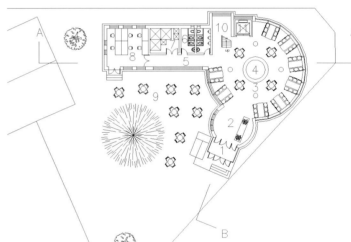

1 Entrance
2 Reception
3 Tea Room
4 Kitchen
5 Lobby
6 Toilets
7 Store Room
8 Office Room
9 External Tea Room
10 Stairs
11 Elevator

一层平面
Ground Floor

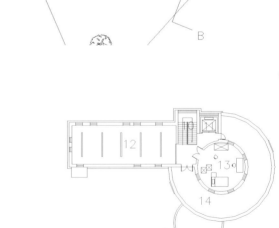

12 Gallery
13 Artist Room
14 External Gallery

二层平面图
First Floor

模型表达
Model Representations

电脑表达
Computer Representations

剖面图

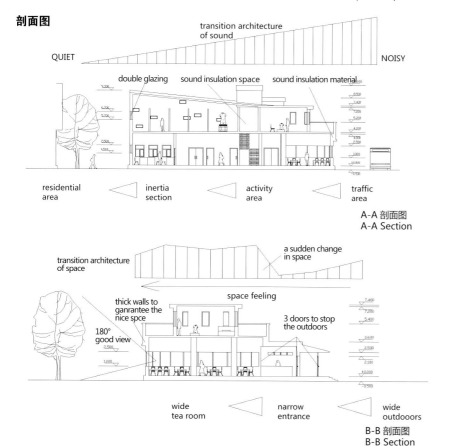

A-A 剖面图
A-A Section

B-B 剖面图
B-B Section

两部分对比鲜明
The two sharply compared parts

屋顶及室外环形画廊
Roof and external loop gallery

室外茶餐厅及大树
External tea room and great tree

方案四

学生姓名：崔浩
The Undergraduate Designer:
Cui Hao

基地策略
Site Strategy

在设计的初始阶段，我们着重分析了中式茶餐厅和西式茶餐厅的区别。随后在Isabel教授的指导下我们挑选出一条关于西式下午茶的文学名言来加深对设计的理解。这条名言也是贯穿整个设计的指导思想。我选择的是一名英国剧作家的话："While there is tea, there is hope."（有茶就有希望）。

基地位于哈尔滨的繁华地带，周围环绕有百脑汇、锦绣科技大厦、哈尔滨工业大学等现代化建筑，饮食以快餐文化为主。快节奏的生活，单调乏味的景色，让人丧失了对美好生活的希望。因此我希望这个英国茶室可以给人们带来新的希望，让人们暂时远离城市的喧嚣，静静地享受生活。

平面及推敲

初期草模
Initial sketch

平面图

1 Water pool
2 Outdoor tearoom
3 Slope
4 Tearoom
5 Reception
6 Toilets
7 Stairs
8 Lift
9 Tea preparing room
10 Store room
11 Gallery
12 Lobby
13 Office room
14 Artist's project room
15 Artist's accomdation
16 Summer bar

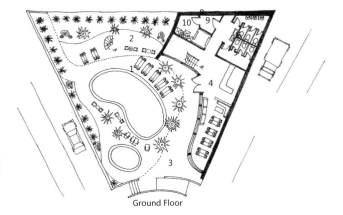

Ground Floor

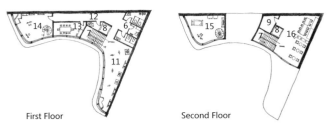

First Floor   Second Floor

剖面图

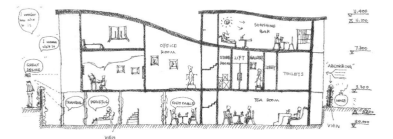

茶室细部表达

入口下沉处理
Underground entrance

建筑底部架空
Empty space

艺术家工作与休闲的场所
Work and leisure for artist

室外茶餐厅及观水平台
Outdoor tea room and water platform

画廊及展览空间
Gallery and the exhibition space

我们认为，作为茶室，精神诉求要远远大于物质诉求。我们向茶室内引入了两个浅水池，通过栽植植物以及铺设草坪，来营造良好的室外环境。

As a tearoom, we focus more on mental demands than physical. We built two waterpool for the outdoor environment, and together with the plants and grass, we can create a better environment for our tearoom.

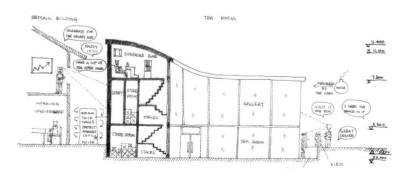

06

Paris Habitat Bridge

Chinese Teachers : Prof. Zhang Shanshan　张姗姗　教授
　　　　　　　　　Prof. Bai Xiaopeng　白小鹏　教授
　　　　　　　　　PhD. Jiang Yilin　蒋伊琳　助教
　　　　　　　　　PhD. Bai Xiaoxia　白晓霞　助教
　　　　　　　　　Assistant MA Zixin　马梓馨　助教

Foreign Teacher : Prof. Eric Dubosc　杜博斯克　教授

Students : Chen Yuting　陈玉婷
　　　　　　Chen Tong　陈　桐
　　　　　　Tao Siyuxiao　陶斯玉潇
　　　　　　Qian Yujie　钱玉洁
　　　　　　Yang Qiaowan　杨巧婉
　　　　　　Lv Tianyi　吕天一
　　　　　　Pan Shuo　潘　硕
　　　　　　Chen Gang　陈　刚
　　　　　　Jing Chenxi　景程曦
　　　　　　Zhang Yunhao　张芸昊
　　　　　　Zhen Qi　甄　琪

2012 Autumn

Academic activity : International Collaborative Teaching and Academic Month
Institutions : Harbin Institute of Technology and ECOLE NATIONALE SUPERIEURE D'ARCHITECTURE DE PARIS LA VILLETTE

巴黎人居桥

合作背景

哈尔滨工业大学与法国巴黎拉维莱特高等建筑学院的教学合作始于2007年，连续五年的短期联合教学课程，探索了国际化联合教学的新模式，获得了很好的教学效果，该课程及指导下的学生连续多次在全国高等学校建筑学专业指导委员会组织的建筑设计教案和教学成果评选中获奖。

此外，由张姗姗教授和白小鹏教授主持的哈尔滨工业大学建筑学院公共建筑与环境研究所于2008年加入了杜博斯克教授主持的国际工作组，该工作室以推进全球建筑学教育合作为基本宗旨，目前成员有法国巴黎拉维莱特建筑学院、里尔建筑学院、清华大学、哈尔滨工业大学、沈阳建筑大学、韩国首尔汉阳大学、韩国晋州庆尚道国立大学等院校。两届研究生合作教学与国际竞赛活动取得了较为理想的成果，相关硕博研究生的合作成果多次获奖。

本次联合设计的外方导师杜博斯克教授是法国国家建筑顾问、法国世界环境保护基金会聘任专家、世界银行建筑节能聘任专家，他主持的项目曾获得都市遗产功勋大奖（法国）、钢结构建筑欧洲奖、金属结构最佳作品奖等多项世界级奖项。杜博斯克教授曾先后受聘于近十所国际建筑院校授课，现任法国巴黎拉维莱特高等建筑学院教授、国际营造工作室主席、法国杜博斯克建筑事务所董事长。

The coorperation between HIT and Ecole d'Architecture de Paris la Villette started since 2007. During the past 5 years, we organized several joint training program, explored international teaching mode. The school work of the course wins a great teaching effect and lots of awards.

In addition, Public Architecture and Environment Institute which is lead by Professor Zhang Shanshan and Bai Xiaopeng was invited to join the International Workshop held by Eric Dubosc. This organization aims to carry forward the global architecture teaching and has more than 20 school members from all over the world. The president of this organization Eric Dubosc is actually the supervisor for this joint training course.

张姗姗　Zhang Shanshan
教授　博士生导师
Prof.　PhD. supervisor

白小鹏　Bai Xiaopeng
教授　硕士生导师
Prof.　Master Instructor

杜博斯克　Prof. Eric Dubosc
法国巴黎拉维莱特建筑学院教授
哈尔滨工业大学客座教授

2009届研究生联合国际联合设计竞赛小组合影

2010届研究生联合国际联合设计竞赛小组合影

本次联合设计的承办方为哈尔滨工业大学建筑学院公共建筑与环境研究所，研究所成立于2006年，是由张姗姗教授和白小鹏教授主持的集理论研究、设计应用研究、实验研究为一体的综合性研究所，于2011年正式更名为公共建筑与环境研究所。研究方向：公共建筑设计理论与方法，包括医院建筑、办公建筑、商业建筑、文化建筑、教育建筑等；建筑行为心理学，包括行为与环境的研究方法、行为心理对城市环境的相互作用、环境行为研究与建筑空间设计等；城市与建筑防灾，包含城市防灾系统研究、建筑防灾系统研究等。

Public architecture & environmental design research institute (PAEI) belongs to Harbin Institute of Technology, which was established in 2006. It has one doctor degree supervisor, one master supervisor, six PhD. students and over twenty graduate students.

There are three main research directions: The first one is public buildings, including hospital buildings, office buildings, commercial buildings, travel buildings, educational buildings, communication and transportation buildings. The second one is environmental behavioral research, including behavior and environment research methods, interaction between urban environment and behavioral psychology, and space design based upon environmental behavior. The third one is city and architecture prevention system against disaster, including urban system research against prevention, architecture fireproof system research and so on.

The PAEI is in charge of many national and provincial fund research projects. And many papers are published on international and national journals and conferences proceedings. Students in PAEI are organized to participate in international competitions, national competitions, and university competitions and received numerous awards.

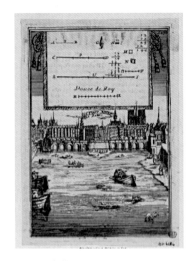

Historical Photos

任务背景

1. 巴黎位于塞纳河中心的两座小岛上，并随着河流桥梁的建设发展起来。
Paris is located between two small islands in Seine River and comes flourished along with the development of lots of bridges.

2. 起先，这些桥梁上有商店、住宅和休闲区域。现今，2012年，为了不久将至的埃菲尔铁塔纪念日，将实现一座通往埃菲尔铁塔的步行桥。
At first, there were some original shops, apartments and spaces for entertainment. In order to celebrate the anniversary of Eiffel Tower, we are planning to build a pedestrian bridge to it in 2012.

3. 这是一座横跨塞纳河的金属桥梁。
It should be a metal bridge across the Seine River.

4. 鉴于临近周边区域已有很多金属桥梁，因此这座似埃菲尔铁塔般的金属桥是可以拆卸的。
Considering that there are already several metal bridges in this area, therefore, this new bridge should be removed just like the Eiffel Tower.

5. 这座桥梁主要承载的建筑有住宅、休闲区域、商店及多功能体，并只供行人通行，但会有为消防车及救护车提供的援助车辆应急通道。
This should be a pedestrian bridge with several kinds of functionalities, such as apartments, shops, leisure time space and so on. However, it must provide emergency channel for vehicles of fire-engine and ambulance.

Historical Photos

Satellite Images

巴黎人居桥梁计划书

1. 建设一座连接两岸，带有商业、休闲区域等功能的人居桥梁。
To build a connection bridge across Seine River with the functionalities of shopping and entertainment.

2. 要求桥梁需横跨230米的河流，每个河堤的宽度为15米。桥连接顶部的河堤（间距是260米）和底部的河堤（间距230米）。应急车道不能受到影响。
This bridge is about 260 meters long and 15 meters wide, which should includes the emergency lane.

3. 商业空间：两层桥梁间包括每间80平方米的商店，带露台的咖啡馆，餐馆归属于商业空间（加减法的表现形式可以是）：800平方米的商店–咖啡馆=1间60平方米，1间100平方米，2间200平方米（不包括露台面积）– 餐馆：2间80平方米，2间100平方米–3间250平方米（含露台）。厨房和餐厅的从属部分占总面积的7%~10%。
Commercial Space: a 80m² store, a café with a large platform, and a restaurant. Your could choose addition or subtraction to make it come true. Subordinate part of the kitchen and dining area should be 7% to 10% of the total.

4. 休闲空间：1间展示巴黎历史的电视会议厅（含大厅和卫生间共300平方米，能容纳200人），1间可容纳200人的夜总会（也是300平方米）。
Leisure Space: A conference hall for the exhibition of Paris History which should accommodates 200 people.

5. 多功能体：45个墙体空间，每间40平方米。可以规划为一层或两层，可做住宅、办公室、手工作坊或画室。
Multiple-function Area: It should includes 45 rooms, each one is 40m². The functionality may be apartment, offices or drawing rooms.

制图要求

1 Site-plan   1:500
2 Plan   1:200
2 Elevation   1:200
1 Section   1:200
3 Details for Construction   1:20
One 3D exterior renderings (or Model)
One 3D interior renderings (or Model)

备注

塞纳河宽：230米
河堤宽：15米
桥墩数目：最多4根
塞纳河的水平面标高：0.00
涨水标高：+3.00
下方桥面的标高：+11.00
两边的桥墩：+8.00
桥下河堤标高：+1.50
顶部河堤标高：+7.00

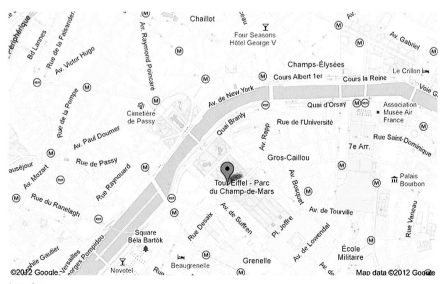

Location

教师和全体学生合影

见面介绍会 Introduction Meeting	
理念与相关技术讲座——外方教授 lecture by Foreign teacher	
开题讨论——全体成员 Introduction of the Program	
学生分组 Make Groups	

Course rhythm

GROUP 1
陈玉婷　陈桐　陶斯玉潇

GROUP 2
钱玉洁　杨巧婉　吕天一

GROUP 3
潘硕　陈刚　景晨曦

GROUP 4
常湘珂　张一帆　陈思宇

GROUP 5
张芸昊　甄琪

在课程的最初，由外籍教授就课程要求和自身的设计理念进行讲解，包括最新的技术、材料的综合应用理论和实例讲解等。课程中期，老师根据学生的理解情况进行跟踪讲解和一些具体的技术设计指导。课程最后，举行学生作业展，老师进行点评并颁发优秀作业证书。
授课流程：
第一天——最新的技术、材料的综合应用理论与实例讲解；第二天至第六天——上午：讲课，下午：作业和辅导，晚上：作业；第七天——上午：作业展览和点评

概念生成
Create concepts

优秀奖：倒影桥

设计评审
Design reviews

概念修正
Concept correction

优秀奖：巴黎雨林

概念深化
Concept deepen

二等奖：三拱桥

在设计评审过程中，老师将本着公平公正的原则对每一组方案给出客观的评价，并评出一个优秀作品。

方案生成
Program generates

技术指导
Technical guidance

优秀奖：塞纳河之上

集中制图
Work out

一等奖：天空之桥

第一组 倒影桥
Group 1 THE REFLECTION BRIDGE

陈玉婷　陈桐　陶斯玉箫
Chen Yuting, Chen Tong, Tao Siyuxiao

设计构思

由分析得出该地轴线关系，将这种轴线关系应用到建筑布局中。将埃菲尔铁塔的形态垂直倒映在水面上形成桥体。主体空间布置在悬吊的柱子上，剩余空间在桥梁端部，形成方案的初步构思。

方案发展

将主要多功能体放在建筑主体部分——支撑桥梁的柱子上，桥上主要放置休闲、商业空间，并可以形成景观很好的室外露台。桥梁中心形成一个活动平台，方便人们之间的交流。

形体生成 Object Creation

The relationship between the building and the Eiffel Tower

场地分析 Site Analysis

总平面图 General Plan

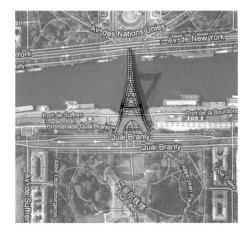

结构分析 Structural Analysis

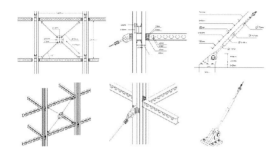

主体结构选用了钢结构。我们尝试以最简化的结构来承受尽可能大的荷载，以弹性结构代替刚性结构来减少外力对结构的损害。

节点草图 Node Sketch

西南立面图 the Southwest Facade

平面及立面图 Plan and Elevation

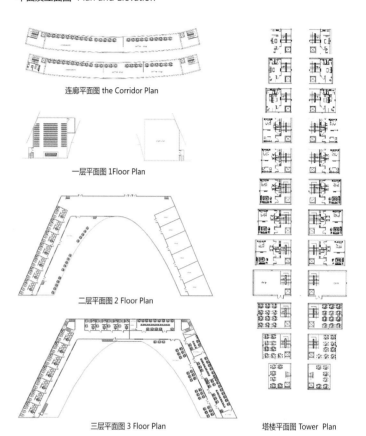

连廊平面图 the Corridor Plan

一层平面图 1 Floor Plan

二层平面图 2 Floor Plan

三层平面图 3 Floor Plan

塔楼平面图 Tower Plan

节点详图 Structure Analysis

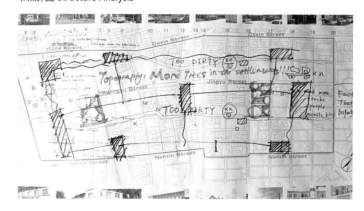

第二组 巴黎雨林
Group 2 THE RAINFOREST OF PARIS

钱玉洁　杨巧婉　吕天一
Qian Yujie, Yang Qiaowan, Lv Tianyi

设计构思

人居式桥梁以桥梁与建筑相结合，赋予桥多种功能空间，促使人们在其上驻留，这与现代大量的为快速交通服务的桥梁形成了鲜明对比。本次设计主题为在横跨塞纳河的耶纳桥的位置建造一座人居桥梁，并作为通往埃菲尔铁塔的步行桥。法国人居桥梁历史悠久，鉴于此，设计构思主要从两点出发：

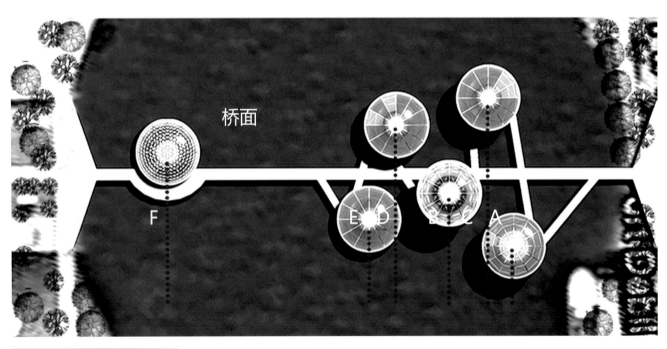

"巴黎雨林"的设计灵感来源于图中塞纳河左岸的国家公园，公园与埃菲尔铁塔隔河相望；公园主轴线直指埃菲尔铁塔，而树木组合成公园平立面上的轴线，故有了"巴黎雨林"的形象拱围着桥面，将轴线感延伸至河对岸的埃菲尔铁塔。

结构以及建筑与环境的关系

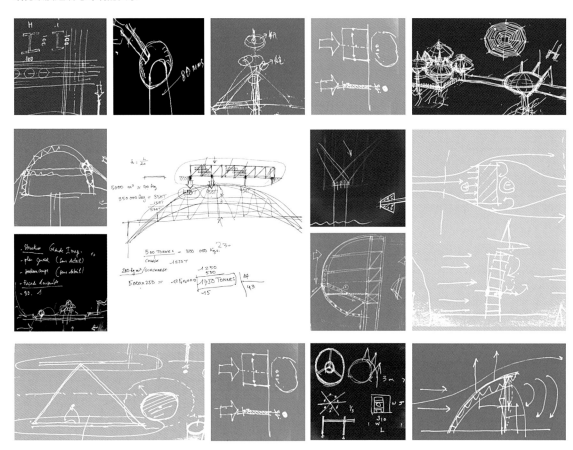

1. 结构方面
桥梁建筑跨度大,承重多,需要具备足够的强度和刚度,我们选用钢材作为承重构件的材料,利用了钢结构的优势。钢结构抗拉性能好,但抗压性能差,在结构设计以及节点构造上进行了细致的推敲和计算。

2. 与环境的关系
桥梁周围存在着河流、堤岸、建筑、广场、绿化、道路等多种要素,它们与桥梁相互依存、互为约束,作为一个整体被感知。因此,桥梁设计应从地区整体出发,与周边要素一起进行综合组织。

1. Structure
Bridge construction loads more, spans larger and needs to have sufficient strength and rigidity, so we choose steel as load-bearing material and take advantages of the steel structure. Steel is good at tensile properties, but poor at compression properties, so we are careful in calculating structural design as well as node structure.

2. Relationship with the environment
Bridges are surrounded with a variety of elements, such as rivers, levees, buildings, plazas, garden, road, which depend yet interact with the bridge, as an entirety to be perceived. Therefore, the bridge design should be designed from an entirety and organize with the surrounding elements.

表现与结构

1. 立面效果图
2. F单体剖面结构图
3. 基座群效果图
4. 不同型号钢材之间以球连接为连接方式
5. 双层网面效果
6. 楼板结构细部
7. A–E单体的基座构造细部
8. F单体的基座构造细部
9. 双层钢网结构
10. 单体结构图

1. Facade
2. Section structure of F monomer
3. Bases
4. The connection between different steel models is ball connection
5. Double layer
6. Floor structure
7. A-E monomer base structure
8. F monomer base structure
9. Double-steel-mesh structure
10. Monomer structure

 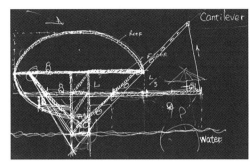  

・单体的上部结构采用工字钢骨架以及双层网壳结构，每层楼板均由钢索和承重骨架相连。
・桥体部分由桥墩处直接伸出的工字钢支架支撑。

・The upper structure of the monomer uses the I-beam skeleton and double-layer reticulated shells.
・Each floor is connected to the cable and the load-bearing skeleton.

・塔身钢架旋转相交，合理传递水平风荷载。
・塔身底部由钢管与底座铰接，并延伸出桥面的支架。

・The tower steel frame intersected rotationally and transmit horizontal wind load reasonably.
・The bottom of the tower hinged by steel pipes and base, extending out of the deck bracket.

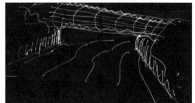

第三组 三拱桥
Group 3 THE ARCH BRIDGE

潘硕　陈刚　景晨曦
Pan Shuo, Chen Gang, Jing Chenxi

设计构思

埃菲尔铁塔作为区域最高点，桥身设计应当提供更多不同的视线角度，同时作为游行人群经过的桥梁，应该提供一定程度的绿化景观与商业休闲要素，因而设计成空中花园式人居桥梁，提供更多的观览空间。

The Eiffel Tower is the highest point of the region, bridge designing should provide more different sight viewing. The bridge as build for protesters should provide some degree of landscape and commercial elements. Our purpose is to design the bridge as a garden which will provide more interesting space to people of Paris and visitors.

我们的设计将桥身分为三段，每段都相互连接而不是封闭的。桥面与桥上建筑分离开来。建筑部分，内部种植树木以形成花园景观。建筑部分由三段巨型月牙拱支撑，桥面部分则是以悬吊的方式固定在月牙拱上。

Our design seperate the bridge into three parts, each one is connected to another. The bridge and the buildings above the bridge is mutually independent. In the buildings we will plant trees to form part of the garden. The supporting construct consists of three seperated giant crescent arches, and the bridge is to overhang right beneath the arches.

总平面图

概念构思 Concept Design

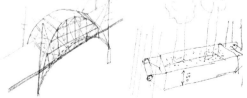

结构设计 Structure Research

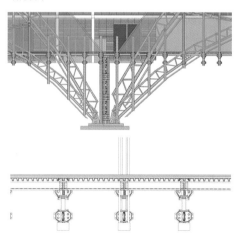

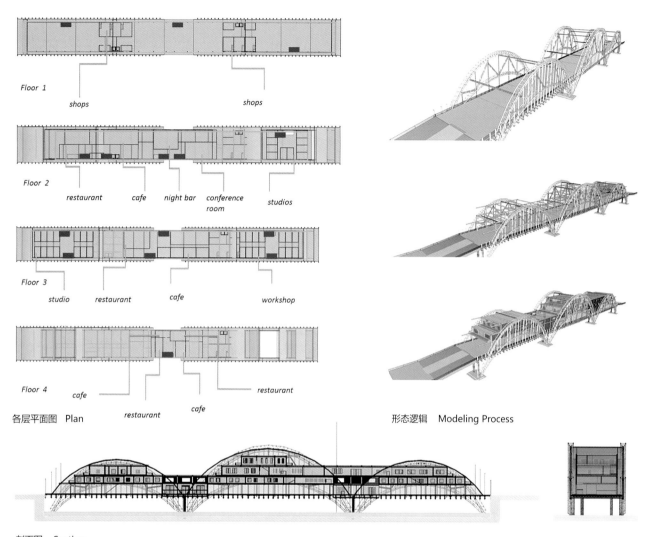

各层平面图　Plan

形态逻辑　Modeling Process

剖面图　Section

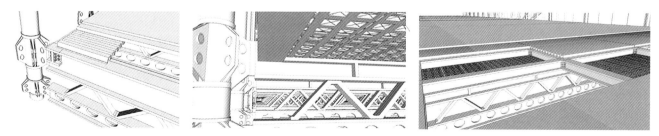

节点详图　Node Detail

第四组 塞纳河之上
Group 4 ICONIC BRIDGE ABOVE SEINE RIVER

常湘珩　张一帆　陈思宇
Chang Xiangzhou, Zhang Yifan, Chen Siyu

设计构思

Located in the Seine river and facing the Effel Tower, this bridge beomes a very important passage for tourists to admire the beautiful scenes. The material of the structure that can be easily assembled is recyclable. At the same time, the bridge offers an opportunity for tourists to fully experience the very details of life in Paris.

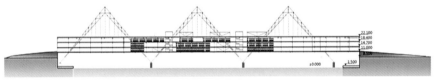

总平面图

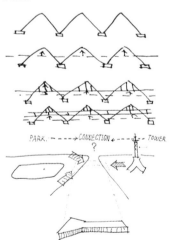

结构分析 Structural Analysis

节点草图 Node Sketch

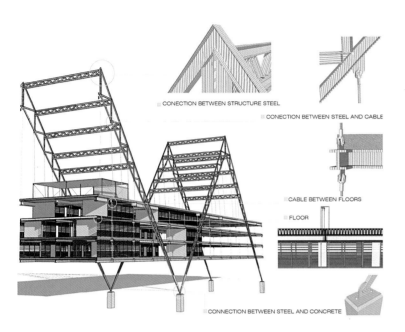

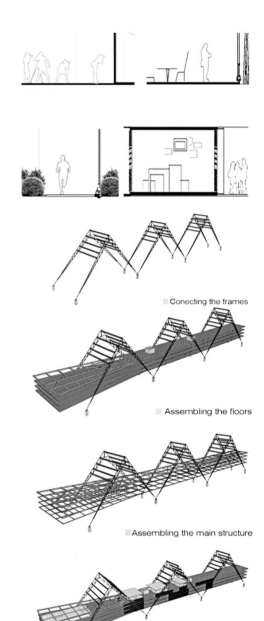

平面图 Plan

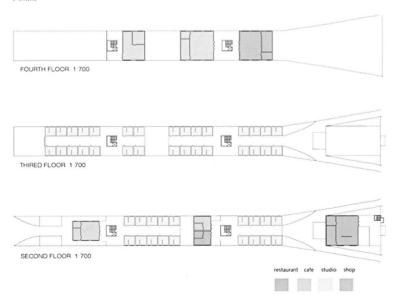

第五组 天空之桥
Group 5 SKY BRIDGE

张芸昊　甄琪
Zhang Yunhao, Zhen Qi

透视效果图 Perspective rendering

如同礼花一般的建筑，静立在埃菲尔铁塔前。钢铁的艺术和精致的节点与同样精美绝伦的埃菲尔铁塔相映成趣。主要结构可拆卸，在庆典过后可以将建筑拆除，单纯作为桥来使用，也是塞纳河上一个钢铁艺术的景观。

The building like fireworks stand in front of Eiffel Tower. Iron and steel art, delicate junctions and the interreflection of the Eiffel Tower set each other off becomes an interest. Main structure is removable. After celebration it can be removed and the construction can be used as a bridge, also, is a steel art landscape on the Seine river .

节点图 Some Junctions   剖面图和立面图 Sectional view & Elevations

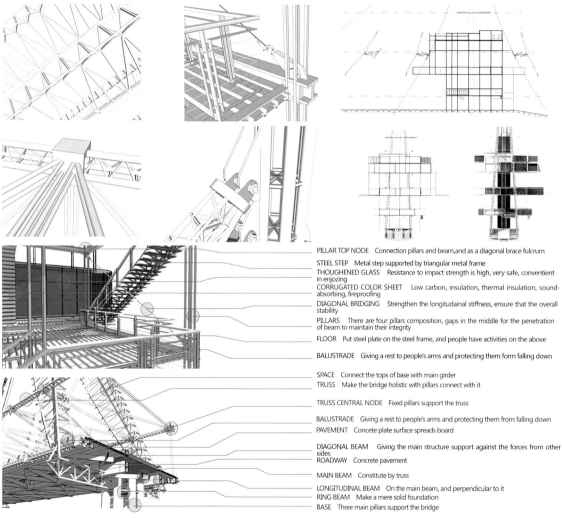

PILLAR TOP NODE   Connection pillars and beam,and as a diagonal brace fulcrum
STEEL STEP   Metal step supported by triangular metal frame
THOUGHENED GLASS   Resistance to impact strength is high, very safe, conventient in enjozing
CORRUGATED COLOR SHEET   Low carbon, insulation, thermal insulation, sound-absorbing, fireproofing
DIAGONAL BRIDGING   Strengthen the longitudainal stiffness, ensure that the overall stability
PILLARS   There are four pillars composition, gaps in the middle for the penetration of beam to maintain their integrity
FLOOR   Put steel plate on the steel frame, and people have activities on the above
BALUSTRADE   Giving a rest to people's arms and protecting them form falling down

SPACE   Connect the tops of base with main girder
TRUSS   Make the bridge holistic with pillars connect with it

TRUSS CENTRAL NODE   Fixed pillars support the truss

BALUSTRADE   Giving a rest to people's arms and protecting them from falling down
PAVEMENT   Concete plate surface spreads board
DIAGONAL BEAM   Giving the main structure support against the forces from other sides
ROADWAY   Concrete pavement
MAIN BEAM   Constitute by truss
LONGITUDINAL BEAM   On the main beam, and perpendicular to it
RING BEAM   Make a mere solid foundation
BASE   Three main pillars support the bridge

节点的设计是大桥和建筑安全的保证。不同部位、不同承重用不同节点。节点的多种多样也成了整栋建筑的点缀，可以让人感受到结构之美，"true-tech"，而非"mis-tech"。

The design of the junctions is the safety assurance of bridge and construction. Different areas, different bearing uses different junctions. Variety of junctions also become the ornament, which make a person experience the beauty of structure. It's "true-tech", rather than "mis-tech".

07

Chess Club for Harbin

Chinese Teachers :	Associate Prof. Han Yanjun	韩衍军　副教授
	Lecturer Chen Yang	陈旸　讲师
	Associate Prof. Wei Dake	卫大可　副教授
Foreign Teacher :	Prof. David Britch	大卫·布里奇　教授
Students :	Sun Jiangshun	孙江顺
	Chen Xixi	陈析浠
	Qian Cong	钱聪
	Wang Xuesong	王雪松
	Xiang Junda	向均达
	Ganzo	刚杂雅
	Lin Zeqian	林泽乾
	Wang Yishuo	王祎硕
	Sun Yuxuan	孙宇璇
	Tan Jiajun	谭佳骏
	Li Lei	李磊
	Han Jiangyue	韩江月
	Ding Weishuang	丁维霜
	Zheng Huijin	郑慧瑾

2012 Autumn

Academic activity : International Collaborative Teaching and Academic Month
Institutions : Harbin Institute of Technology and The University of Sheffield

哈尔滨国际象棋俱乐部

任务背景

"景观不在现代运动的议程上,除了作为卫生或娱乐的关注,还作为绿地或'tapis verte'。曾经是花园与城市设计的主要点——审美与象征方面,基本上被现代运动丢弃了。这在很大程度上解释了景观建筑在近代处于建筑的从属地位,现代花园或景观的传统很少存在了。"

以上评论来自Stuart Wrede题为《取自古典白话中的景观与建筑》(1986,ISBN 0-08478-0678-2)的文章。

"Landscape was not on the agenda of the Modern Movement, except as a sanitary or recreational concern, as a greenbelt or 'tapis verte' The Aesthetic and symbolic dimension, which had traditionally been a central concern of garden and landscape design, was essentially discarded by the Modern Movement in favor of utilitarian concerns. This largely accounted for the fact that landscape architecture in modern times has found itself in a subordinate position to architecture and that a vital modern garden or landscape tradition can hardly be said to exist."
Article by Stuart Wrede titled Landscape and Architecture Classical andVernacular by Asplund taken from A splund published by Rizzoli 1986 ISBN 0-08478 -0678-2

这是一个对建筑与景观之间关系的探索,毕竟这两个学科之间具有相同的基本元素——空间。

This is an exploration of the relationship between Architecture and Landscape, bringing together two acts of making, after all both disciplines work with the same basic element – SPACE.

这所设定建在哈尔滨的象棋俱乐部是一个拥有约70名成员的国际象棋小社团,成员来自城市的各个角落:大学学者、当地居民和车间工人,他们每周聚会3次并定期举行比赛。

他们要求:
· 小型会议室:可以承办委员会会议,配有厨房、洗手间等设施
· 灵活空间:可以在一次象棋比赛中容纳30人
· 小礼堂:作为当地学校的拓展空间时可以扩大一倍
· 小型咖啡书吧
· 可以下棋的室外空间:一个重要的元素,在俱乐部试图使象棋游戏更流行时提供室外场地
· 建筑必须符合DDA标准

The Proposed Chess Club for Harbin is a small chess society with about 70 members coming from all over the city, academics from the universities, local residents and shop workers. They meet 3 times a week and hold regular tournaments.

They require:
· A small meeting room for committee meetings with a kitchen and toilet facility.
· Flexible space capable of holding chess tournaments of about 30 players at a time.
· A small lecture theatre which can double up as an outreach space for local schools.
· Small book shop cafe.
· An outside space for playing chess, this is an important element, as the club wants to make the game of chess more accessible.
· The building should be fully DDA compliant.

合作背景

谢菲尔德大学建筑学院学风扎实严谨，有着优良的传统，誉满欧洲。哈工大建筑学院学生本着认真严谨的学习精神开始了这次联合设计之旅。学院正在向"研究性设计教学"转变，联合设计为学院注入了新的活力，意义非凡。

School of Architecture, University of Sheffield School has a rigorous style of study and a fine tradition, which is of great fame in Europe. Then our students of Harbin Institute of Technology began this rigorous study of design. Our college is changing to build a "research design teaching", so this joint design injected new vitality, and is of great significance.

在这次同RABA成员David教授的联合教学中，David特有的英式幽默深刻地感染了每个人，课堂气氛活跃。同时，David的教学方法也深深启发了我们。每堂课先由同学们简要介绍自己的方案，再分组授课的教学方法，既保证了同学们之间的交流和了解，也充分发挥了小组讨论的作用，很好地提高了教学效率。课程启动前期，我们同David通过邮件往来几经修改任务书，最终敲定基地并明确了教学任务。这个过程在以后的联合教学中可以借鉴并完善。

In this joint teaching with RABA member Professor David, David inspired everyone with his peculiar British humor and actived classroom's atmosphere. Meanwhile, David's teaching method also inspired us deeply. Each class starts with a brief introduction of all the students on their own programs, and then split the class. This ensures communication between students, and also gives opportunity to the role of group discussions, so it's a good way to improve teaching efficiency. We start early course via e-mail exchanges with David, and finally we made the teaching task clearly. We can learn and improve a lot through this process.

全体同学和老师们的合影
All the students and teachers

· **Monday**

Site Analysis I will briefly describe the objectives of the project we will then have a series of workshops, on the quality of the site. This is an opportunity for students to describe how they interpret the site. These observations are to be backed up by models, drawings and photographs. Describe the quality of the chosen site using both empirical tools such as mapping, surveys, figure ground analysis urban patterns, as well as more personal interpretations of the site using semi abstract representations You will then be asked to produce a site strategy, describing how the site will be occupied by the use of the built form and its relationship to the landscape.

· **Tuesday**

Site strategies will be examined and discussed. Following the workshop you will then be asked to produce a abstract 1:500 model which will attempt to bring together the understanding of the site the volumetric requirements of the brief and the architectural narrative that each student wants to explore.

· **Wednesday**

1:500 abstract model will be examined and discussed. You will then be asked to produce some preliminary plans and sections where the emphasis is on the relationship between the landscape and the building, expressing the fluidity of the spaces and the connectivity between inside and outside space.

· **Thursday**

Plans and sections will be examined and discussions will take place. You will then be asked to produce a 1:100 model, which will explore the relationship between structural volume and envelope and how these synthesize into something architectural. You will then be asked to produce a section through the building and landscape.

· **Friday**

1:100 Models will be examined and tested against the ideas being explored, its important at this stage not to forget what the project is about, at the root of the project is the idea of an architectural ideas or narrative sitting within a larger context.

· **Saturday**

Endgame You will be expected to complete the project pulling all the week efforts into a clear 1:100 model, some plans and sections showing the landscape and building, at 1:200 and a series of perspectives that move us through the landscape to the building.

· **Sunday**

Review the work

· 周一

基地分析
从基地的层面简述我们这个小组的设计目标。同时，同学们有机会阐述各自对基地的理解。这些对基地的观察会通过同学们的模型、草图和照片反映出来。既可以通过已有经验，如地图、调查、数据，也可以通过个人感受，如一些半抽象的东西来感受基地和城市肌理。接下来要求做一个基地策略，描述基地是如何被建造满的以及和景观的关系。

· 周二

基地策略
以小组为单位，制作1:500的基地草模，这个草模试图使同学们综合理解基地对建筑体量的要求及基地建筑性的表述。

· 周三

1:500草模
设计表达一些初步的平面和剖面，侧重点是景观与建筑的关系，拟在表现空间的流动性和内外的连通性。

· 周四

平面与立面
评估与讨论平面与立面。制作一个1:100的模型，模型要求考虑建筑的结构、体量、表皮等要素以及它们怎么以建筑的方式结合表达出来。要求做一个建筑与景观的剖面。

· 周五

1:100模型
评估与检验1:100模型，模型要求基于上次课后深入的想法。这个模型必须体现方案自身的目的，而这个方案必须基于一个建筑上的想法或者置身于一个更大的环境中。

· 周六

最终方案
最后要求将之前一周的努力做成一个1:100的模型，一些1:200的能够表现空间与景观的平面和剖面以及一系列展示建筑与景观的透视图。

· 周日

展示成果

方案一

孙江顺　Sun Jiangshun

基地位于中国哈尔滨，位于哈尔滨工业大学建筑馆和居民区（20世纪80年建造）之间。基地周遭有历史建筑，缺少公共空间，活动的主要人群为居民以及学生。

Ths site is located in Harbin, China, which is between HIT architectural department hall and residential areas. Historical buildings exist around the site, which is lack of public space. The major crowd are students and residents.

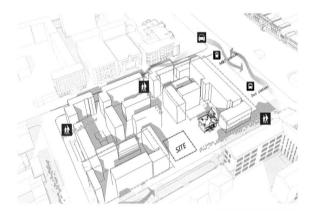

基地周边的交通流线
Transportation Streamline

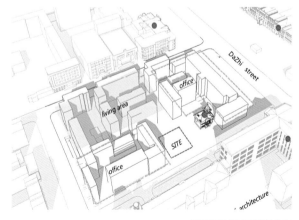

基地周边保护建筑的分布
Historical Building distribution

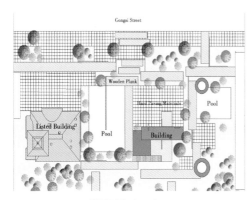

总平面图 Site Plan

1. 本设计想要营造便于周遭居民、学生活动的公共空间。
2. 将基地中的记忆元素（烟囱、水塔）重新组合，使建筑和原有城市记忆有机结合。
3. 想要使居民楼和建筑馆之间有一个良好的过渡。
4. 营造一个适合各个年龄层棋牌娱乐的建筑单体。
5. 为哈尔滨爱好棋类活动的人群提供一个长期使用的场所。

1. To create a public space for students and resident .
2. To restore the memories of historic building and old structures (chimney, water tower).
3. To create connections between residential buildings and architectural department of HIT.
4. To create a chess space for all ages people of all ages near the site.
5. To create a long-term space for chess enthusiasts in Harbin.

First Floor Plan

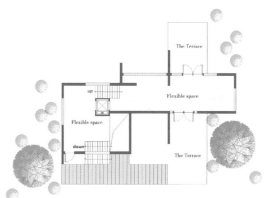

Second Floor Plan

Northeast Facade

Northwest Facade

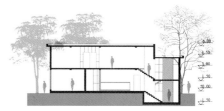

A-A Section

方案二

孙宇璇　Sun Yuxuan

树下之弈——哈尔滨国际象棋俱乐部设计
Chess under the tree-Chess club design in Harbin

一期：场地策略，David要求我们深入地分析场地，原本已经调研过的基地在他的带领之下又重走一次，他对周边的水塔、古树、古建筑都有深入的解读。确定保护建筑的位置和场地中的古树，让新老建筑形成对话，同时，留出穿梭车辆和行人的边界道路用于限定场地。入口的确定考虑建筑学院的学生和原本的居民。

First: site strategy . We are supposed to make a deeper analysis of the site including the location of the water tower, old trees and historic building. The conversation between new and old buildings and the circulation of the pedestrians are well taken into consideration.

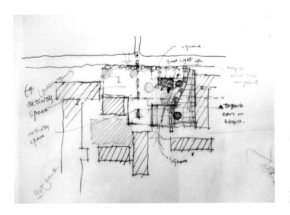
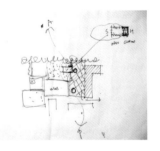

上图：David 手绘
左图：场地策略

二期：初步勾画平面草图，确定了场地的布置。我开始思考建筑和基地之中的要素如何更加和谐地对话。在基地中有几棵古树可形成一个小中心，建筑呈环抱状，后面水塔、烟囱一侧是硬质的铺地。建筑正是一个面朝草地，背靠道路的"阴阳脸"，这体现在平面上就是开敞和封闭空间的划分。

Second: a rough plan and arrangement of the whole site. An old tree is centered by the building . I begin to consider how to arrange the elements in order to make them more harmonious. Open and enclosure space are devided and demonstrated in the plan.

三期：确定平面，模型制作。最后的平面确定为两个相互平行的内弯的弧形，外延的道路作为建筑的延伸。朝向西面的部分被植被分割开来，由于水塔和车行道路的缘故，空间也比较封闭，在面向保护建筑的一侧，空间开敞，营造了可以树下对弈、草木中游戏的场景。倾斜的屋顶充分注重与保护建筑的协调尊重。

Third: final plans and model making. There are two arcs that are parallel one another. Roads are extended from the main buiding. The western part in comparatively enclosed space. Instead the western part create the atmosphere that people can enjoy their leisure time playing chess there and other recreation. The leaning roof shows respect to historical buildings.

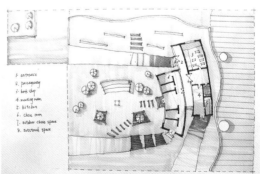

上图：平面
左图：手绘透视

方案三

王雪松　Wang Xuesong

获得任务书时第一感觉就是在一个极大的地段上做一个非常小、非常简单的建筑。
所以这次设计的主要任务就不是简单地做好一个建筑,而是去解决建筑与环境的关系,通过
场地的设计来丰富基地的空间,与基地的原有肌理统合进行设计。

Obtained when the first impression is the mission statement in a great location to do a very small very simple construction.
Therefore, the main task of this design is not simply to do a building, but to resolve the relationship between architecture and the environment through the design of the site to enrich the base of space. And to enrich the space of site through the desigh, combining with the original textile at the same time.

设计初期确定建筑靠近古树设置,这样可做到与有趣景观的互动,又通过与古建筑保持距离来尊重古建。

It was determined that the building be placed close to the tree, for that interesting landscape can be accessible.

场地布局

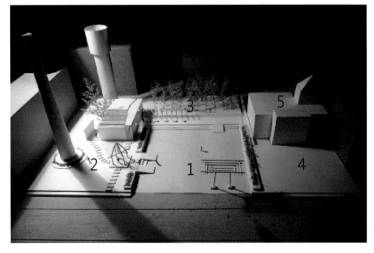

1. 开放活动场地
2. 俱乐部内院
3. 树林休息空间
4. 保护建筑门前广场
5. 保护建筑院落

节点透视

主要空间透视　　　　　　　　　　　　细节景观处理

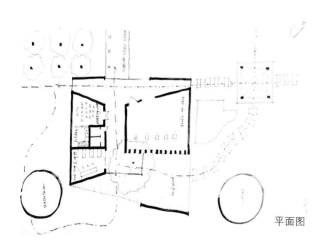

平面图

在建筑的设计上，考虑了建筑与场地设计的结合，建筑内部设计了一条通道，使其不阻挡居民区到公路的流线，同时另一条通道连接了古建与后院的流线。两条互相垂直的通道划分了建筑的内部空间。两条通道的交汇处成为建筑的核心。建筑设计初期过多地考虑了建筑如何与古树、水塔、烟囱发生联系，建筑面向古树一侧开敞，面向古建一侧封闭。

We take the combination of building and site into consideration, and design two passages inner the building. Vertical one another, they decide the initial space. During the initial stage, we think more of the way how building talks to the enviornment.

方案四

向钧达　Xiang Junda

老城区中的国际象棋俱乐部
CHESS CLUB IN OLD CITY AREA

当我走进这个基地的时候,我就知道设计将要面临的主要课题是建筑、环境与城市记忆的关系。徘徊在基地之中,无数问题在我的大脑中浮现:在网络高度发达的今天,人们为什么要走出家门,来到这里下棋?为什么这个基地有着一种难以说清的魅力?我在设计中应该如何传承这种魅力?我的建筑要如何同历史建筑发生对话?这些早已失去功能价值的烟囱、水塔,其存在价值是什么?

随着这些问题的浮现,无数建筑形象在我脑中闪过:平顶的、坡顶的、历史的、现代的、华丽的、极简的……然而站立于这两座丰碑一样的砖砌建筑物旁,任何形式都显得那么不堪一击。于是我看到了这些老朽的建筑身上带着的来自大工业时代的骄傲,它们自尊心极强,它们需要被后来的人们关注、凝视。于是我的建筑从三个方向指向三个历史建筑,并把一颗古树拥入怀中。故事在耳边诉说似乎能更加深刻,于是建筑与建筑几乎亲吻在了一起,在这个象棋俱乐部中,我们能看到每一块砖的粗糙纹理,听得见每一块砖期待被关注的声音……

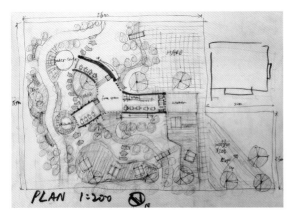

平面设计草图

概念模型

When I get into this site, I know immediately that the most important issue I have to face is the relationship among architecture, environment and city history. Wondering in the site, many questions get into my mind: what's the most important spirit of this site? How can I convey that spirit through my design? What can I make use of with these seemly useless chimney and water tower?

I suddenly feel that every brick in these industrial heritages is trying to say something, they are the pride of that industrial era, their voices need to be heard… so the structure of my design is to stretch to face these three these industrial heritages, so that in the chess club we can even see the texture of each brick clearly.

没有平立剖面，也没有分析，我想一个好的设计应该是像在讲述一个故事。我觉得模型是最能表达一个建筑性格的媒介，在模型里可以看到材料，可以感受空间，所以我选择用一个精致的模型来传达我的故事。

There is no need to show any plan, section and elevation of this building. I believe that handmade model is the most suitable media to describe the character of an architectture.

上右、上左：成果模型
中左：成果模型
下左：空间意向草图
下右：成果模型

当我们体验一个建筑的时候是看不到任何图纸的，也是没有任何人为我们介绍的。建筑师想要说的一切只能透过建筑这个实体"不言自明"。所以对于这个联合设计，我想我得到最大的收获就是用模型去推敲方案吧。

方案五

陈析浠　Chen Xixi

本设计方案的基地位于建筑学院侧门对面的生活社区内，内部有一座哈尔滨保护建筑，周围有住宅和部门机构。针对复杂的人群结构和建筑环境，本方案采用简洁的设计手法，用一块连续并起伏的平面将基地分为两个部分，南侧为以原有的烟囱和水塔还有保护树木为视觉重点的休憩空间，北侧为以历史保护建筑为主题的开放空间。这块连续的曲面既是建筑的屋顶与表皮，也是北侧空间的限定要素。

The site is in the residential neighborhood within a historical building. Faced with complexed population and environment, I use a simple strategy with a continuous and undulating surface dividing the site into two parts. The south part includes an old chimney, a historical water tower and protected trees. The north side is an open space at the theme of the historic building. The continuous surface is not only the roof of the building with the epidermis, but also the defining elements of the north side space.

处于这样的环境中，既要考虑为社区居民提供舒适方便的活动场地，也要为城市提供开放的交流场所。

In this kind of environment, it is necessary to consider to provide a comfortable and convenient playgrounds for community residents. In the other hand, it provides an places for the city.

基地分析

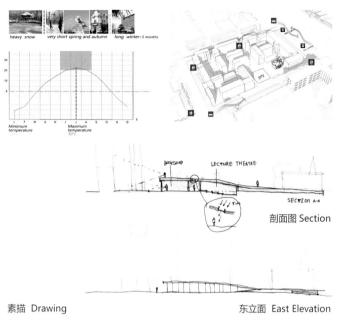

剖面图 Section

素描 Drawing　　　　　　　　东立面 East Elevation

设计过程

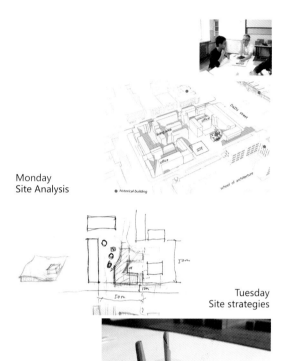

Monday
Site Analysis

Tuesday
Site strategies

Wednesday
1:500 abstract model

Thursday
Plans & Sections

Friday
1:100 Model

方案六

丁维霜　Catherine Ding

折廊　WINDING GALLERY

本设计基于棋子的运动方式以及不同人群进入场地的过程中心理的变化和需求的研究。场地景观的设计和布置，与人群进入场地后的心理需求相对应，通过高差的变化和景观的塑造，营造出三个不同主题的场地趣味空间。

According to the analysis below, I make my landscapes meet the need of the psychological demands of different people. By putting the site in different levels and rebuilding the landscape, I divide the site into 3 interesting parts.

■ 棋子的运动方式　The movement of a piece

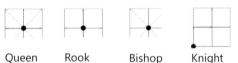

Queen　Rook　Bishop　Knight　Pawn

不同功能棋子的行走方式，就像不同人群在不同心理需求时需要的行走方式。
The movement of a piece is just like the different spatial demands of different people.

■ 不同人群从远离场地到进入场地活动的心理变化
The change in mental when go to the site

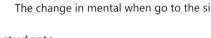

■ 空间心理与空间需求相对应
The demands in the site and what the space look

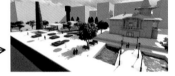

Part 1 is for the sightseeing

relax　sit　play　see the scenery

free　complex and interesting

人群需要可以自由行走和有趣味的休闲空间。

Part 2 is a slope for rests and playing

play chess　see others play chess　take some activities

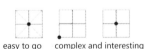

free　easy to go　complex and interesting

同时各个地区的活动情况要易于观察、参与和到达。

easy to find　easy to go

cross　meet and chat　play chess or see others play chess

人群进入场地目的性强，需要便于识别方向的场地环境，各地区活动情况易于观察和到达。

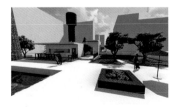

Part 3 is for the chess club

折廊

建筑设计的概念来源于中国古典园林中的廊式空间，人们在曲折的廊式空间中自由地行走与停留，并且取得良好的多向的景观。建筑的造型与观景口使得三个原有建筑间产生对话。

The concept of the design comes from the passage in Chinese traditional garden. People can enjoy themselves when they cross or stay in the passage; what is more, they can see great landscapes outside from many directions.

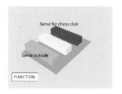  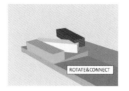

以满足建筑功能的合理性为出发点，通过功能体块的移动、旋转形成了最终的建筑形态。变形后的建筑空间给予了建筑两个内向性的庭院和檐下的半室外灰空间，并且在建筑内视线上将三个原有建筑联系起来。

Based on the reasonable function of the club, I put the 3 parts into different levels, then, rotate and connect them. As a result, 2 gardens for club are formed and there are some disheartened spaces as well. The club connects the 3 spots through the sight.

广场中部的草坡让远离道路的市民生活区更加宁静、私密，草坡上可以进行各种各样丰富多彩的市民活动，不在广场停留的居民也可以从下沉通道无干扰地穿越广场。由于草坡的高差，使得建筑呈现位于不同水平线上的两个出入口，同时满足对外和对内人群的需要。

The grassy slope in the middle of the square makes the area for residents more quiet and private. People can enjoy their colourful daily lives in the grassy slope, while people who don't want to stay in the square can cross the square by the underground path. Because of the Level difference, there are 2 main entrance in different levels. Both of them can meet the need of the people outside and residents.

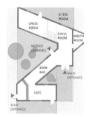
Site Plan

Chess Room 3

Chess Room 2

Cafe

Bookbar and Cafe

Bookbar

Meeting Room

Chess Room 1

方案七

钱聪 Qian Cong
Moving "Chesspieces"

David:"那个烟囱去哪儿了?"
Money:"我把它推倒了,然后用砖做了可供房子移动的轨道。"
David:"棋盘上所有的东西都能移动吗?"
Money:"是的,除了水和树。"
David:"Money,你是个疯狂的女人。"

David : Where is the chimney ?
Money : I put it down, and build the rails for the movement of these houses.
David : Can everything on the chessboard move ?
Money : Yes, except the water and the trees.
David : Money, you are a strange women.

设计理念

设计的主旨在于"活动"二字,将场地设计成一个大块的国际象棋棋盘和一个真人版对弈场地。棋盘的划分由轨道形成,呈现的小块地区分成草地、土地、铺地和水潭,地块的深浅就像棋盘的黑白。建筑的设置并未集中在一起,而是将面积分散,形成多个小建筑体块,通过体块的移动组合实现大小面积的功能转换,比赛时便聚集在一起,自由对弈时便可滑动至自己心仪的处所,同时,轨道上的座椅、凉棚、对弈墙、桥廊等均可移动,可谓"时移景异"。

The keyword of this design is "Moving". We design the site into a large chess board and a live version of chess venue. Chessboard division formed by the track, showing the small area into grassland, land, paving and pool, plots shades like black and white checkerboard. Architectural settings are not together, forming multiple small building body mass, which move through the combination of the size of the area to achieve functional transformation. They come together when the game is on, while sliding to where you like when it's free to play. Meanwhile, the track chairs, pergola, chess walls, bridge gallery, etc. can be moved. Described as "scenes in different time."

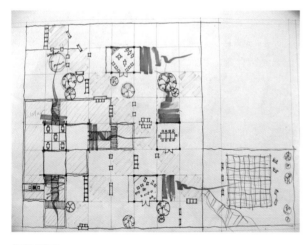

场地平面

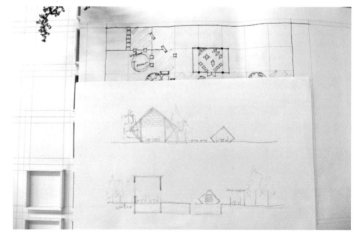

关于立面的设计,休息部分和对弈部分采用两种不同的模式。两个休息区是将矩形体块加以翻转,切去底部,使其活泼;四个对弈区是6×6×6的立方体,都设有可移动、可拆卸的活动墙。

As for facade design, the rest part and the chess part are in two different modes. Two rest areas are rectangular bodies to be flipped, with the bottom of the block cut, making themselves lively. Chess areas are 6×6×6 cubes with movable walls.

真人对弈场
Live chess games

场地与轨道
Going with the track

在很宽阔的场地上建面积很小的建筑，我们应该怎样做？方案设计着眼于周围的景观设计，可移动的桥、对弈设施，使各处小景观更好地相互融合。

Building a house in a very wide place. What should we do? This design focuses on the surrounding landscape, removable bridge, chess facilities, making the entire small landscape better integration.

移动对弈墙
Moving chess wall

自由的道路
Freedom of the road

树下与廊下
Trees and porch

博弈邀老宅
Game invites old houses

设计最初的想法是在基地内种上薄草坪，任由人们自由地在其上行走。人们会按照平日去市场和车站的习惯，踩出两条合理的道路。

The original idea was planting grass at the base of a thin lawn, allowing people to freely walk on it. People will go to the market and the station with the usual custom, trod reasonable ways.

在林中草旁
Grass in the forest beside

移动的建筑
Moving buildings

08 From Concrete to Abstract
From Abstract to Concrete

Chinese Teachers :	Prof. Lu Shiliang	陆诗亮　副教授
	Lecturer Liu Ying	刘　莹　讲　师
Foreign Teacher :	Prof. Lorenzo. Barrionuevo	罗伦索·巴里奥努艾奥　教　授
Students :	Jian Rui	菅　睿
	Han Xu	韩　旭
	Dong Mingxuan	董名轩
	Zhou Fan	周　凡
	Xiang Yuchen	项雨晨
	Lu Shuran	陆书然
	Liu Shengze	刘圣泽
	Song Chaoyue	宋超越
	Wang Yuxuan	王雨萱
	Xiong Yexin	熊叶昕
	Luo Jie	罗　杰
	Zhang Zhanou	张占欧
	Ha Yuanyuan	哈元源
	Huang Xi	黄　茜

2012 Autumn

Academic activity : International Collaborative Teaching and Acacemic Month
Institutions : Harbin Institute of Technology

从具体到抽象，从抽象到具体

合作背景

哈尔滨工业大学自2010年以来邀请了大量外籍注册建筑师指导联合设计，这些短期联合设计课程引起了众多本科生和研究生的极大兴趣，取得了很好的教学效果。与国外资深建筑师联合设计教学有助于培养能够适应国际前沿和建筑学科领域发展需要，熟悉国际设计规则和惯例，具有创新精神、创造能力和较强沟通能力的国际化卓越工程师。

Since 2010, Harbin Institute of Technology has invited large numbers of foreign registered architects to instruct design studio. These studios has aroused great interest among numerous undergraduate and graduate students and obtained very good teaching effects. Collaborative design education together with foreign senior architects could contribute to train excellent engineers who could adapt international forefront and the needs of architecture discipline development, familiar with international design rules and regulations, have innovative spirit and creative ability, and strong communication skills.

本次联合设计的外方导师巴里奥努艾奥教授是巴塞罗那ARQTEL城市、建筑、室内设计及工程公司的合伙人及联合创建人，城市运动建筑合作伙伴及联合创建人，西班牙高级建筑师协会理事会成员（CSCAE），加泰罗尼亚建筑师联盟会员（COAC）。巴里奥努艾奥教授曾先后受聘于西班牙马德里圣巴布罗大学、巴库阿塞拜疆建设与建筑大学、中国东南大学等多所国际建筑院校。自2011年以来，在哈尔滨工业大学指导可持续建筑与城市设计课程。

In this studio, foreign advisor Prof. Barrionuevo is the partner and co-founder of ARQTEL urban, architecture, interior design and engineering firm, architecture partner and co-founder of City Movement, member of The Spanish senior architects Association (CSCAE), member of Catalonia union of Architects (COAC). Prof. Barrionuevo has also been teaching in many international architecture schools such as Universidad CEU San Pablo, Madrid, Spain; Construction and Architecture University, Baku, The Republic of Azerbaijan; Southeast University, Nanjing, China, etc. Since 2011 he instructed several studios and seminars on sustainable architecture and urban design at Harbin Institute of Technology.

陆诗亮　Lu Shiliang
副教授　硕士生导师
Pro. Graduate tutor

刘莹　Liu Ying
讲师　博士
Lecturer Doctor

罗伦索·巴里奥努艾奥
Lorenzo Barrionuevo
教授　博士
Prof. PhD.

任务背景

建筑的设计及建造过程总是在一定条件下展开的，并经历着从抽象到具体，抑或从具体到抽象，乃至不断反复的过程。

Architectural design and construction process are always carried out under certain conditions, and experience a continuous iterative process from either abstract to concrete or from concrete to abstract, and even goes round in circles.

所谓"一定的条件"，可能关乎具体的气候、环境、地形等。

The so-called "certain conditions" may be related to specific climate, environment, topography, etc.

所谓"具体"和"抽象"，可能关乎光线、气味、质感、材料、构造、建造、思考、方法、几何、关系、图形、功能等。

While "concrete" and "abstract" may be related to lighting, smell, texture, material, configuration, fabrication, thinking, methods, geometry, relations, graphics, function, and so on.

本次联合设计课的主旨在于探索可持续建筑设计与技术，经过与同学们沟通，外教同意采用天作杯大学生设计竞赛任务书，各组同学自选地段并自拟设计题目。

The aim of this studio is to explore sustainable architectural design and technology. By communicate with all students, Prof. Barrionuevo agree to use the assignment of "Team Zero" National Undergraduate Architectural Design Competition. Every group of students could choose their site location and subject by themselves.

设计任务书

主题：从具体到抽象、从抽象到具体
Theme: From concrete to abstract, from abstract to concrete

竞赛委员会为设计者提供了一个略抽象的基地，在绵延坡度为10°、18m×18m投影的场地正中上方，还有一块6m×6m的空中水平基地，建筑的可建造范围将被限定（任意小于但不超过）在如图虚线所示的两个立方体范围内，上部立方体为6m（长）×6m（宽）×6m（高），下部立方体为18m（长）×18m（宽）×9m（高）。场地朝向自定，气候可根据参赛者所在地的具体气候或假想为寒冷的、炎热的、多雨的、干旱的、多风的、阳光充足的……环境可以是丛林、草地、沙滩……功能安排，在上部立方体范围内须为居住，下部立方体范围内功能不限（居住或其他），但上下功能须能并置成立。上述朝向、气候、环境不限，但必须自我设定。

The Competition Committee provides designers a slightly notional site. The site is in the middle of the 18m × 18m square projection field with continuous slope of 10 degrees, and a 6m x 6m square horizontal site floating in the air. The project should be constructed within (any less but not more than) the two cubes, in which the upper cube is 6m (length) x 6m (width) × 6m (height), and the lower cube is 18m (length)× 18m (width) x 9m (height). Designer can decide site orientation and climate conditions, such as cold, hot, rainy, dry, windy, sunny, etc. The environment can be a jungle, grassland, beach ⋯ The function of upper cube must be residential, while the lower cube is not limited (residence or otherwise), still the functions up and down should hold up together. The above orientation, climate and environment are not limited but have to be set up.

提交图面：总平面、平面、剖面、立面、透视图或模型照片以及200字以内的设计说明。
用纸：以上各图及设计说明均需安排在一张A1尺寸大小（600mm×840mm）的图纸之内，不裱图板，图面表现方法不限。

Submit Drawings: master plan, plans, sections, facades, perspective drawings or photographs of physical models, as well as design specification no more than 200 words.
The above drawings and design specifications should be arranged on an A1 size (600mm × 840mm) drawing paper, don't mount on the drawing board, drawing expressions are not limited.

| 概念生成 | 概念修正 | 概念深化 | 方案生成 | 技术指导 | 集中制图 |
| Create concepts | Concept correction | Concept deepen | Program generates | Technical guidance | Work out |

学生分组

第一组：

菅睿　　　　　　　　　韩旭
Jian Rui　　　　　　　Han Xu

第二组：

董名轩　　　　　　　　周凡
Dong Mingxuan　　　　Zhou Fan

第三组：

陆书然　　　　　　　　刘圣泽
Lu Shuran　　　　　　Liu Shengze

第四组：

宋超越　　　　　　　　王雨萱
Song Chaoyue　　　　Wang Yuxuan

第五组：

熊叶昕　　　　　　　　张占欧
Xiong Yexin　　　　　Zhang Zhanou

第六组：

项雨晨　　　　　　　　罗杰
Xiang Yuchen　　　　Luo Jie

第七组：

哈元源　　　　　　　　黄茜
Ha Yuanyuan　　　　　Huang Xi

过程照片

感想

王雨萱 Wang Yuxuan：

联合设计是我来到哈工大之后做的第一个设计，作为交流生，我很感谢学校给我们提供这样一个机会，能够面对面与外教老师和其他班级同学一起进行交流、学习。罗伦索教授为人随和、守时，对我们的设计悉心指点，但由于时间的关系，最终的方案设计还有很多地方需要完善和深入。

This project was the first one after I came to HIT. As an exchange student. I appreciated school to offer the opportunity for us,to communicate with foreign teachers face to face. Professor Lorenzo is easy-going and punctual, and then,he gave many advice to our project. But, due to time constraint, there are many places where the scheme design needs to be improved further.

宋超越 Song Chaoyue：

我要感谢lorenzo Barrionuevo教授、刘莹老师、陆诗亮老师以及我的队友陪伴我度过了一段短暂而美好的设计周期，我很享受期间寓学于乐的过程，很欣慰最终满满一车的成果，可持续发展的设计要求让我深深地意识到建筑师对社会、对环境肩负的责任。做有责任感的建筑师，我已在路上。

I want to express my appreciation to professor lorenzo. barrionuevo, Miss liu, Mr lu, and my partner. With their accompany, I experienced a short but beautiful design cycle. I enjoyed puting study with entertainment during the progress, and satisfied with the abundant accomplishment .

砂时计 Hourglass

GROUP 1 菅睿 韩旭

砂时计，又名沙漏，古代的计时仪器，寓意为时间的流逝。倾斜的沙漏，会有一部分沙子永远不会流下，象征着那些经典的艺术在岁月的流逝中被保留下来，而沙漏的形状也代表了上部分生活和下部分艺术展示的关系，虽然不尽相同，却总是有联系。

Hourglass, also known as the hourglass, is the ancient chronometric instruments, and the implication for the passage of time. Tilt of the hourglass, there will be a part of sand will never down, is a symbol of the classic art are preserved in the years passed, and the shape of an hourglass also represents the relationship between life and art show. Despite the difference, there always be a certain relation.

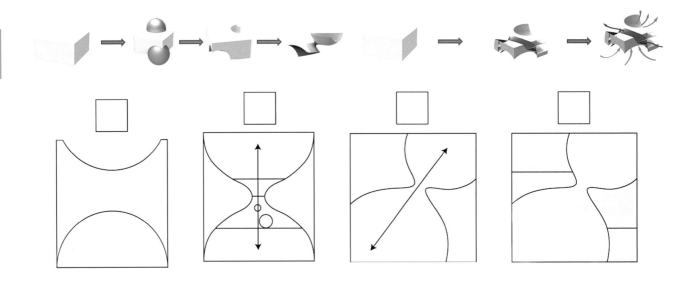

基本使用功能：上部分生活，下部分艺术展示

以沙漏的形式使上下两部分产生联系

由于上方方块的存在，使得中轴垂直沙漏的上半部空间消极，调节中轴使其倾斜

将下部分作为演示区，下方方块的其余部分则作为工作室和展示区

立面图

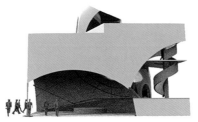 

交通流线图

平面

展厅　舞台

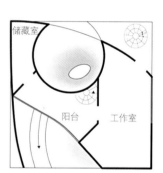

储藏室　阳台　工作室

开放空间

生活区　卫生间

From 38°C To −38°C

GROUP 2　董名轩　周凡

概念

在松花江畔，在冬夏温差近70°C的哈尔滨，夏季的流水和冬季的冰雪似乎是自然环境的全部。大自然赋予了水神奇的变化。我们正在试图建造一个与这一美妙过程相关的建筑。

In Songhua River of Harbin there are 70°C difference in temperature between winter and summer.
The summer water and winter snow and ice.
All seems to be the natural environment.
Nature gives the magical change to water.
We are trying to build a building associated with this great process.

平面和剖面

空间

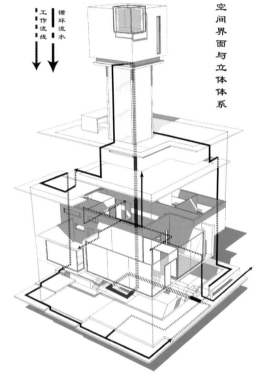

立面

轩

夏季水位上升，下部建筑随水位上浮。建筑被水幕包围，成为水屋，在炎炎夏日中为室内降温。

中部楼梯基础深入地下，支持上部居住空间。下部建筑由砌体基础支撑，在水中产生浮力使下部空间可随江水涨落上下浮动。

随着水位变化，下部建筑与中央楼梯相对位置变化，建筑本身成为水位浮标，上部居室的采光窗将在夜间成为灯塔，指引航向。

上部居住空间与下部工作空间通过楼梯连接，中央楼梯被水覆盖与江水融为一体，及覆盖下部屋身的水幕随季节涨落。

上部方块看似漂浮空中，仿佛消失一般。而实际上，浮动的是下部建筑。

这便是：满眼风波多闪烁，看山恰似走来迎。仔细看山山不动，是船行。

总平面图

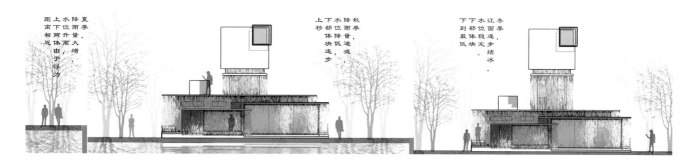

Houses in river 屋美价廉
GROUP 3 刘圣泽 陆书然

近30年以来，中国经济腾飞，大量土地被征用以修建新建筑，房价也随之一路飙升，高不可攀。出现了两极分化的现象，有人拥有不止一套房子，而大部分老百姓可能永远都住不上"高档"住宅。

For nearly 30 years, China's economic boom. A lot of land has been expropriated to build new buildings, and house price is soaring, unattainable. Appeared the phenomenon of polarization, some people have more than a house, and most people could never live in "upscale" houses.

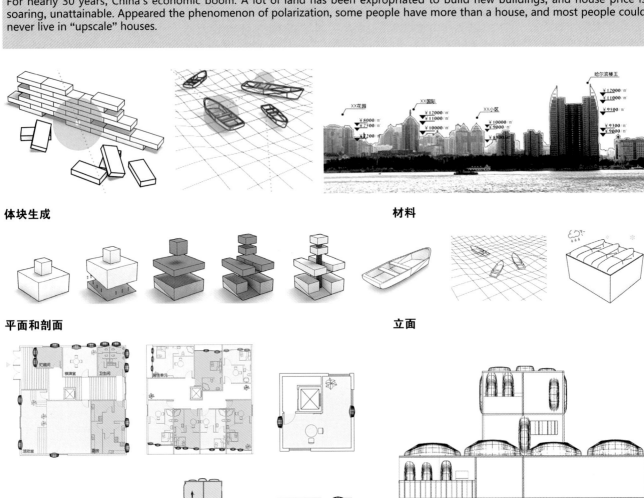

体块生成

材料

平面和剖面

立面

效果表达

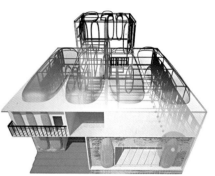

And-or-Not "与" 或 "非"

GROUP 4　王雨萱　宋超越

"从具体到抽象，从抽象到具体"，面对这一课题，我们从两个角度进行思考：如何定义环境？如何定义建筑？旅行，作为现代人的一种生活方式，尤其被越来越多的年轻人所推崇。他们或搭车，或骑行，或徒步，终点只有前方。
空间是建筑的灵魂，元素是空间的符号。限定性元素繁衍既定性建筑，非限定性元素派生多功能空间。当空间脱离对环境和生活的抽象认知，建筑便展现出一个整体、鲜活、复杂的世界。站，在旅途中，在生活中。

"From specific to abstract, from abstract to specific". In the face of this subject, we though from two aspects: how to definite environment? And how to definite architecture?Travelling, coming to be a style for people to live, especially for the young, they go self-driving tour, cycling or hiking, and they are always on the way.
Space is the soul of the building, while element is the symbol of space. Finite element reproductive both qualitative building, while none finite element derived multi-functional space. When space break away from the abstract cognition of the environment and the life,building will show a whole, fresh and complex world. Station, is on the way, is in our life.

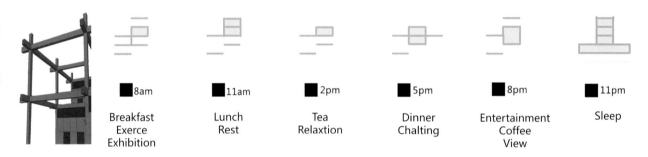

8am	11am	2pm	5pm	8pm	11pm
Breakfast Exerce Exhibition	Lunch Rest	Tea Relaxtion	Dinner Chalting	Entertainment Coffee View	Sleep

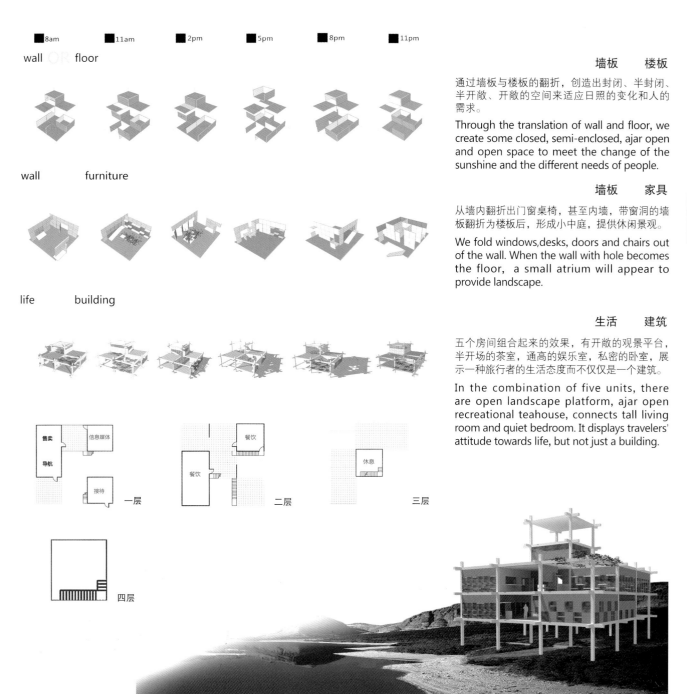

DIY-ENERGY Housing

GROUP 5　张占欧　熊叶昕

滑雪是较为复杂的运动，滑雪前需要一些简单的准备活动。然而很多人并不注意这一点，在滑雪前不做热身，这样会使关节、韧带与肌肉在静止与固定的模式中，没经过任何预热而马上进入应急状态，这种强制应急埋下受伤的隐患，轻者会造成肌肉拉伤，重者甚至会骨折。每年进入雪季，医院都会接到很多滑雪受伤的患者，危害不容忽视。所以我们在滑雪场设计了一个促进游客热身运动，并通过运动产生能量以及得到优惠住房的快捷旅社，并结合当地气候进行可持续设计。

Skiing is a more complex sport, before skiing a few simple warm-up is essential. However, many people do not pay attention to this point, they does not warm-up before skiing. This would make ligaments and muscles in still fixed pattern immediately enter the emergency state without any warm-up. This would planted the hidden danger of injury. People will get muscle strain to some degree, seriously even fractures. Hospitals will receive a lot of ski injuries patients into the snow season every year. And the harm can not be ignored. So we design a holiday inn in ski field to encourage visitors to do warm-up to get more light and electric energy and have a dis-count cost for room.

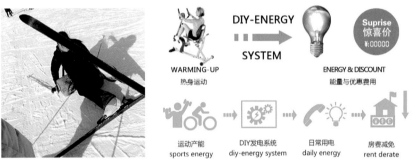

节点构造

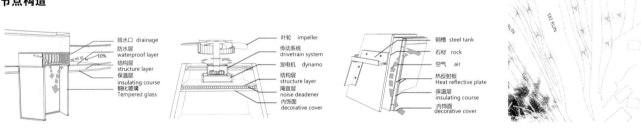

循环系统

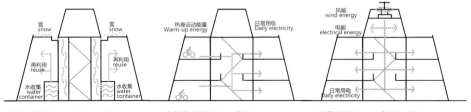

雪水利用 Use of Snow Water　　生物能利用 Use of Bioenergy　　风能利用 Use of Wind Energy　　积雪利用 Use of Snow Cover

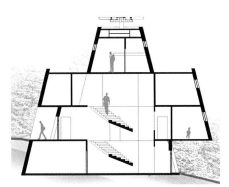
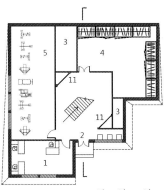
First Floor Plan

Second Floor Plan

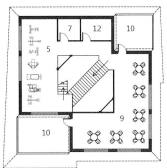
Third Floor Plan

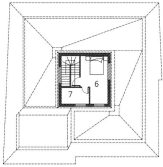
Top Floor Plan

平面图

1. 办公室 Office
2. 休息区 Waiting Area
3. 蓄水设备 Water Storage
4. 储藏室 Storage
5. 健身房 Sports Area
6. 客房 Room
7. 卫生间 Bathroom
8. 接待厅 Reception
9. 咖啡厅 Cafe
10. 室外平台 Outside Platform
11. 玻璃景观排水管 Drainage & Scenery Glass Tube
12. 厨房 Kitchen

GREEN CUBE "绿"方体十二人工作室

GROUP 6 项雨晨 罗杰

本方案位于中国北方寒地地区，寒温带大陆性季风气候。方案设计为12名建筑学学生及其导师提供宿舍及工作室。方案从一颗抽象的温度洋葱出发，利用基地内高差，建筑主体在规定空间内获得了上、中、下三部分分区，动静功能予以区分，同时形成可设定不同温度的独立采暖空间，旨在提高建筑的可持续性。同时，设计使学生卧室部分可动，极端气候下聚拢以减少窗地比及体形系数，气候温和时散开，以利采光通风。

This project is located in the northern cold regions, cool continental monsoon climate. Scheme design for 12 architecture students and mentors provide dormitory and workshop. Solution from an abstract temperature onion, using the base elevation difference in building main body within the allotted space won 3 three parts partition, activity function to distinguish, at the same time can set different temperature independent heating space formation, aimed at improving building sustainability. At the same time, the design part of the movable causes the student to the bedroom, under extreme weather window to reduce the ratio and shape factor, scatter when climate is mild, and daylighting is ventilated.

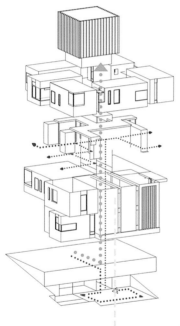

剖面和立面

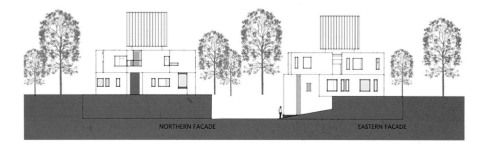

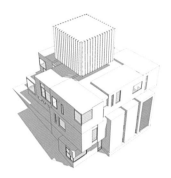
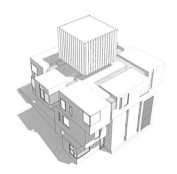

平面

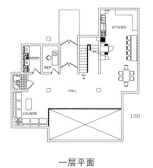

| 地下一层 | 一层平面 | 二层平面 | 三层平面 |